Rammani Adhikari

Energy Detection Based Spectrum Sensing in Cognitive Radio

AF153182

Rammani Adhikari

Energy Detection Based Spectrum Sensing in Cognitive Radio

GlobeEdit

Impressum / Imprint
Bibliografische Information der Deutschen Nationalbibliothek: Die Deutsche Nationalbibliothek verzeichnet diese Publikation in der Deutschen Nationalbibliografie; detaillierte bibliografische Daten sind im Internet über http://dnb.d-nb.de abrufbar.
Alle in diesem Buch genannten Marken und Produktnamen unterliegen warenzeichen-, marken- oder patentrechtlichem Schutz bzw. sind Warenzeichen oder eingetragene Warenzeichen der jeweiligen Inhaber. Die Wiedergabe von Marken, Produktnamen, Gebrauchsnamen, Handelsnamen, Warenbezeichnungen u.s.w. in diesem Werk berechtigt auch ohne besondere Kennzeichnung nicht zu der Annahme, dass solche Namen im Sinne der Warenzeichen- und Markenschutzgesetzgebung als frei zu betrachten wären und daher von jedermann benutzt werden dürften.

Bibliographic information published by the Deutsche Nationalbibliothek: The Deutsche Nationalbibliothek lists this publication in the Deutsche Nationalbibliografie; detailed bibliographic data are available in the Internet at http://dnb.d-nb.de.
Any brand names and product names mentioned in this book are subject to trademark, brand or patent protection and are trademarks or registered trademarks of their respective holders. The use of brand names, product names, common names, trade names, product descriptions etc. even without a particular marking in this work is in no way to be construed to mean that such names may be regarded as unrestricted in respect of trademark and brand protection legislation and could thus be used by anyone.

Coverbild / Cover image: www.ingimage.com

Verlag / Publisher:
GlobeEdit
ist ein Imprint der / is a trademark of
OmniScriptum GmbH & Co. KG
Heinrich-Böcking-Str. 6-8, 66121 Saarbrücken, Deutschland / Germany
Email: info@globeedit.com

Herstellung: siehe letzte Seite /
Printed at: see last page
ISBN: 978-3-639-73018-0

Zugl. / Approved by: Kathmandu University

Abstract

This book discusses the performance study of different types of energy detection schemes used in Cognitive Radio. In this book, the conventional single threshold based spectrum sensing algorithm has been simulated and its performance in terms of probability of detection has been proved to be lower than that of the double threshold based algorithm, at low SNR conditions. The performance of double-threshold energy detection has also been also compared in this book based on two parameters: probability of collision and the probability of spectrum unavailability. All simulations are performed in Matlab.

Acknowledgement

I would like to thank my wife Laxmi Acharya. Without her love and support I would not have made it through the many late nights and countless hours of these works.

Table of Contents

List of Figures

List of Tables

1 Introduction

1.1 Background

There has been a huge growth in wireless users and applications in recent years. The reason behind this is rapid advancement in the wireless technologies. Over the past decade, cellular and personal communications have become the fastest growing segments of the telecommunications services. The demand for wireless services has been absolutely exploding. Several digital technologies have been introduced and enhance wireless communications by adding features and services such as facsimile and data transmission and new call handling features. A new vision of ubiquitous personal communication systems has been introduced with the start of twenty first century. Furthermore, many new wireless communication services and capabilities using the wideband CDMA technology are being defined by the industry worldwide. With the higher data transmission speed and the new services such as simultaneous voice and data, multimedia and location services, applications of the new wireless technologies to the biomedical field have become practical. Moreover, many research works on wireless sensor networks (WSN), next generation network services, telemedicine, smart home appliances are in crucial phase.

Table 1-1: Range of frequencies in U.S. for most common wireless technologies [20]

Applications	Frequency or Frequency Range
AM broadcast	530kHz – 1.7 MHz
Broadcast television	54 – 88, 174 – 216, 470 – 698MHz
FM broadcast	88 – 108 MHz
Cell phones	750, 850, 1700, 1950, 2100 MHz
GHS (non - military)	1.5 GHz
Satellite radio	2.3 GHz
Wireless computer networks	2.4 and 5.8 GHz
Satellite TV	12 GHz
Fixed point – to – point links	1 – 90 GHz

Wireless networks are characterized by a fixed spectrum assignment policy around the world under the regulations of regulatory bodies; for instance, Federal Communication Commission (FCC) in United States in private sector, Office of Communications (OFCOM) in United Kingdom, National Telecommunications and Information Administration(NTIA) again in U.S. at the government level and Nepal Telecom Authority (NTA) in Nepal. The major services and their radio frequency allocation in United States have been shown in the Table 1.1. From it, we see that almost all common wireless applications presently utilize spectrum below 100 GHz. The vast majority of applications are at frequencies below 30 GHz.

Studies showed that large portion of the assigned spectrum is used sporadically and geographical variations in the utilization of assigned spectrum ranges from 15% to 85% with a high variance in time. The sporadic usage of spectrum is illustrated in figure below:

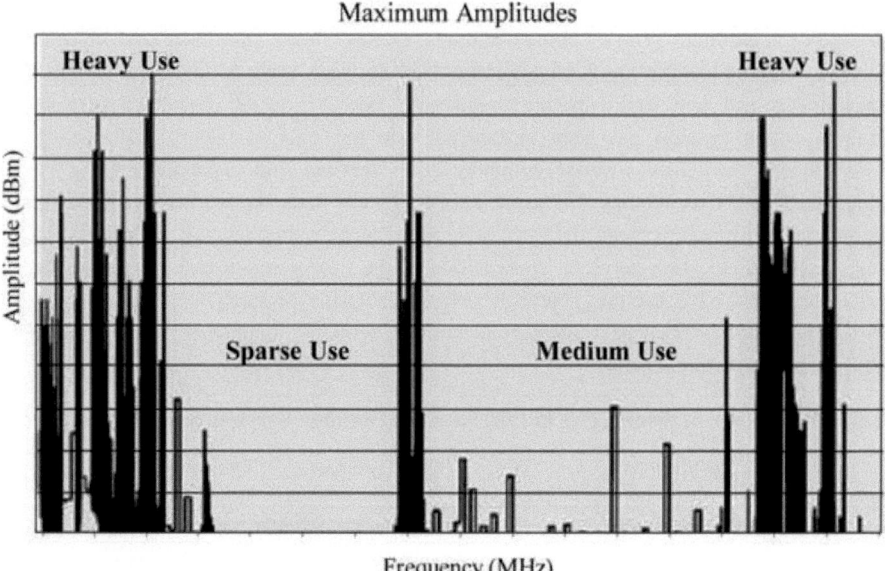

Figure 1.1: Spectrum utilization [20]

From the Figure 1.2, the spectrums in the lower and higher frequencies are heavily used whereas certain frequencies are medium in usage and some frequencies are not used or sparse in usage. [20]The unused spectrums are called spectrum holes or white spaces. Further, the result released by FCC indicates that for 90% of the time many licensed frequency band remain unused. For example, the average utilization of the licensed spectrum for television (TV) broadcast was found to be 14% in 2004, and this number keeps decreasing every year because of the commencement of cable TV and satellite TV. This suggests that it is not an actual spectral shortage but rather the inefficient spectral usage. As demand on the number of users and their data rates steadily increase, it is important to enable efficient access and use of the radio spectrum. The studies of Federal Communications Commission (FCC) spectrum policy task force has illustrated that in certain licensed bands below 3 GHz, the usage of spectrum is very low at any given

2

location and time. This study group suggested that TV bands operating in UHF/VHF are less significantly used and many upcoming mobile applications and wireless applications can be introduced on this band with some technology. The limited available spectrum in the sense that they are already licensed but not efficiently used, therefore; new applications are competing for the very little spectrum that is left unlicensed or in some cases are not getting spectrum. This scenario defines a new communication paradigm. This new paradigm should allow users to exploit the existing wireless spectrum in a shared manner.

The spectrum regulatory bodies in different regions are facing challenges to accommodate the upcoming wireless communication users and applications within the limited electromagnetic spectrum. Besides these, there are data networks in local and personal area, which are called WLAN and WPAN, respectively, working in the license free band called industrial scientific medical (ISM) frequency bands. These bands allow multiple wireless applications to coexist in a given shared environment. The most widespread systems in the 2.4 GHz ISM band are IEEE 802.11 and Bluetooth. Thus, with the proliferation of new wireless technologies and the increasing bandwidth demands, unlicensed ISM spectrums for exclusive use to some extent by increasing infrastructures has become possible. These bands are already crowded with the introduction of increased utilization of wireless technologies operating in this frequency band such as Cordless Phones, Wireless personal area network, HIPERLAN, microwave ovens, wireless video camera, wireless game controller, Zigbee, wiMAX. The co - channel and adjacent channel interference to some extent and the interferences from other non 802.11 applications cause significant loss of throughput of 802.11. In some countries, the numbers of smart phones are more than their population. Then, it is likely that huge interference among those users is intended. Therefore this band is not the permanent and long-term solution to address the tremendously coming wireless applications and services.

In a given country, for example, there are ten users using fixed frequency bandwidth and there is no more frequency slot left. The new wireless devices manufactures or service providers want to use the frequency for their devices. Since, there are no frequencies left, they cannot operate their devices. As explained earlier, not all the frequency users simultaneously occupy the fixed spectrum. It is logical that those unused licensed spectrum can be used by other service providers under the condition when licensed users are not using the bands. Dynamic spectrum access (DSA) defines such a scenario. The DSA therefore is a new paradigm capable of using the electromagnetic spectrum dynamically by pooling up the frequency. This new paradigm is also referred to as NeXt Generation (xG) Networks.

DSA can be best viewed with the help of the Figure 1.2. It can be seen that the band is

3

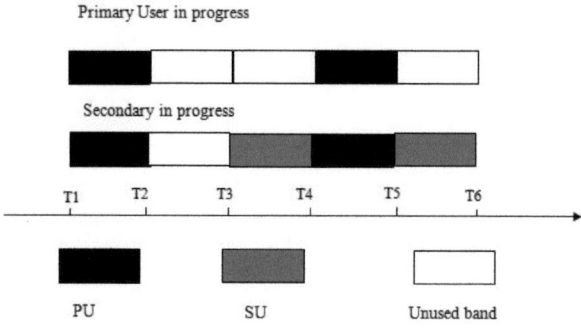

Figure 1.2: Concept of cognitive radio

used for time intervals: t1 to t2, t3 to t4 and t4 to t5 by primary user. The band is observed vacant for time intervals: t2 to t3 and t5 to t6. The secondary may make use of it for those vacant intervals. Again the SU has to evacuate the channel when the primary wanted to use it. The secondary user should be intelligent enough that it should vacate the spectrum when the primary users want to use its usual frequency spectrum. However, the electronic radio chip used in the traditional communication devices are quite dumb in the sense that they do only things according to the hardware circuits of the chip designed for the purpose. The new paradigm shift in wireless communication, especially in the radio technology with the use of software defined radios, enables intelligence to be embedded such that the radios can think and act accordingly. This provides a development platform to a new wireless technology known as cognitive radios. The cognitive radios are defined by the specific functionalities in order to think and act, learning the radio environment and gathering intelligence with the corresponding decision making processes. The embedded intelligence in the radios is then used to perform efficient communications by optimizing the usage of the scare radio resources, such as the radio spectrum.

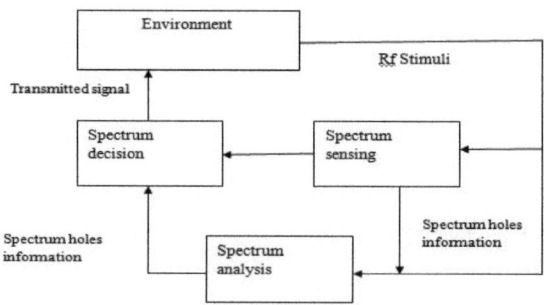

Figure 1.3: Cognitive radio cycle

The overview concept of cognitive radio has been given by Mitola through cognitive radio cycle. The cognitive radio cycle proposed by him is depicted in the Figure 1.3.

According to the cognitive radio cycle a cognitive radio monitors spectrum bands, captures their information, and then detect the spectrum spaces or spectrum holes or white spaces. The characteristics of the spectrum holes that are detected through spectrum sensing are estimated. The appropriate spectrum holes are chosen according to the user's requirement. Once the best bands are determined, the communication can be performed over this spectrum bands. This concept was then adopted by Haykin by defining the respective physical layer communications and signal processing associated with it. This then lead to in depth treatment of cognitive radio research around the world, giving a huge number of research papers and the articles produced on the very same topic.

There has been no globally adopted official/formal definition for cognitive radios as yet; however different definitions are presented in the literature as well as by the radio regulatory authorities around the world. Some of the known definitions of cognitive radio are summarized as follows:

" A really smart radio that would be self RF and user aware, and that would include language technology and machine vision along with a lot of high fidelity knowledge of the radio environment [1]."

-Mitola

"Cognitive radio is an intelligent wireless communication system that is aware of its surrounding environment(i.e.; outside world) and uses the methodology of understanding by building to learn from the environment and adapt its internal states to statistical variations in the incoming RF stimuli by making corresponding changes in certain operating parameters(e.g.; transmit power, carrier frequency, and modulation strategy) in real time, with two primary objectives in mind: Highly reliable communications whenever and wherever needed; Efficient utilization of the radio spectrum [3]."

-Simon Haykin

"Cognitive radio system (CRS): A radio system employing technology that allows the system to obtain knowledge of its operation and geographical environment, established policies and its internal sate; to dynamically and autonomously adjust its operational parameters and protocols according to its obtained knowledge in order to achieve predefined objectives; and to learn from the results obtained."

- ITU – R

"Cognitive radio (design paradigm -1): An approach to wireless engineering wherein the radio, radio network, or wireless system is endowed with awareness, reason, and agency to intelligently adapt operational aspects of the radio, radio network, or wireless system."

5

"A type of radio in which communication systems are aware of their environment and internal state and can make decisions about their radio operating behavior based on that information and predefined objectives [18]."

- IEEE (DYSPAN)

"A radio system employing technology that allows the system to obtain knowledge of its operational and geographical environment, established policies and its internal state; to dynamically and automatically adjust its operational parameters and protocols according to its obtained knowledge in order to achieve predefined objectives; and to learn from the results obtained."

- ETSI RRS

It has defined CR as "A radio or system that senses its operational electromagnetic environment and can dynamically and autonomously adjust its radio operating parameter to modify system operation, such as maximize throughput, mitigate interference, facilitate interoperability across secondary markets."

- FCC

Different peoples and organizations defined CR in different ways. All of them pointed to the same core idea- bringing intelligence to radios or embedding intelligence into radios that could then learn, adopt and react accordingly. The licensed allotted/legitimate user with own fixed spectrum is defined as primary user or incumbent user. The other user is secondary user which uses the vacant bands or white spaces in an opportunistic manner without interfering on the communication of the primary user. It means the secondary user adjusts its carrier frequency; transmit power, modulation scheme, coding etc. to make best use of the white spaces with good quality of service. As soon as the primary comes for its service, the secondary either goes away or masks itself without causing any interference to the primary user. Therefore, the secondary needs to know whether the primary is present or not before using that spectrum; which makes the spectrum sensing a challenge in the implementation of cognitive radio. Spectrum sensing is the ability to be aware about the underutilized bands both in temporal and spatial domain. These underutilized or fallow bands are referred to as spectrum holes or white space. The performance of the CR depends on the degree of detection of vacant bands because the interference can be lowered to negligible amount with good detection.

1.2 Overview of spectrum sensing

There have been different detection schemes proposed in the context of cognitive radio applications including Energy Detection (ED), Matched Filtering, Feature based sensing, and other sensing techniques (covariance based methods and Eigen - value based methods). Different techniques serve different purposes based on their advantages and disadvantages [6].

1.2.1 Energy based spectrum sensing

It is the simplest method for detecting primary users in the environment in a blind manner. It is computationally efficient and could also be used conveniently with analog and digital signals at the RF/IF stages or at the base band. It has a well-known drawback in the detection performance when the noise variance is unknown to the sensing node. Based on the knowledge of noise power, the detection performance of the energy detector can be improved.

1.2.2 Cyclostationary feature based spectrum sensing

In wireless communications, the transmitted signals show very strong Cyclostationary features (sine wave carriers, pulse trains, repeating spreading, hoping sequences or cyclic prefixes) based on the modulation type, carrier frequency, and data rate, especially when excess bandwidth is utilized. Therefore, identifying the unique set of features of particular radio signal for a given wireless access system can be used to detect the system based on the Cyclostationary analysis at the cognitive node. For a sufficient number of samples, this method can perform better than the energy based detection method when the Cyclostationary features are properly identified. However, the main drawback with this method is the complexity associated with it and the requirement for a large sample set for better estimation and precision of the features in the frequency domain.

1.2.3 Matched filter based spectrum sensing

For matched filter based spectrum sensing a complete knowledge of the primary user signal is required; for instance the modulation format, data rate, carrier frequency, pulse shape etc. should be known in advance by the secondary. This gives better probability compared to previously stated methods using the energy detector and the Cyclostationary feature based detector. But, it requires complete signal information and needs to perform the entire receiver operations (like synchronization, demodulation, etc.) to detect the signal.

1.2.4 Other methods

The covariance based methods rely on the fact that the covariance of wireless signals and the additive noise component are generally different. The difference is therefore used to detect the presence of a wireless signal by distinguishing from the noise signal. The eigenvalue-based method for spectrum sensing and detection is again based on the computation of the covariance matrix of the sensed signal.

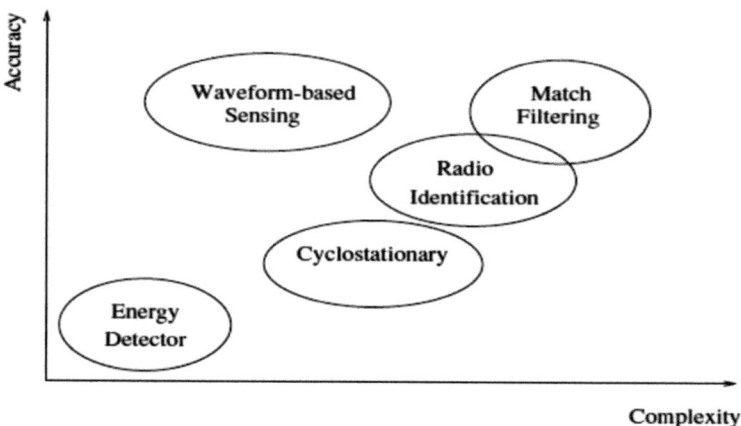

Figure 1.4: Main sensing methods in terms of their sensing accuracies and complexities

1.3   Comparative study of sensing methods

The general comparison of sensing methods discussed so far is presented in the Figure 1 - 5. Waveform based sensing and Matched filter based are more robust than energy detector and cyclostationarity based methods because of the coherent processing that comes from using deterministic signal component. However, there should be a priori information about the primary user's characteristics and primary users should transmit known patterns or pilots. Even though matched filter based method and Cyclostationary feature based sensing method is found to outperform energy based sensing method, the requirement of knowledge of the primary signal characteristics at the expense of long observation time make them less suitable for spectrum sensing in cognitive radio environment, where knowledge of the primary user signal is usually unavailable. Thus, in cognitive radio applications the energy based sensing is the simplest and is optimal.

Spectrum sensing by cognitive radios can be conducted either individually or cooperatively. Recently, the efficacy of cooperative spectrum sensing has gained a great deal of attention. Even though, cooperative spectrum sensing offers several disadvantages as compared to non-cooperative, due to randomness of appearance of PUs, it is extremely difficult to achieve fast and smooth spectrum transition leading to limited interference to PUs and performance degradation.

8

1.4 Statement of Problem

Spectrum sensing is the most basic functionality of cognitive radio, the accuracy of which in turn determines the performance of cognitive radio networks. The objective of spectrum sensing is to sense the vacant bands with no primary user activity. The simplest sensing method, often used in cognitive radio application is the energy based detection method. There are many techniques under energy based detection such as the Single threshold algorithm; Double threshold algorithm; Adaptive threshold algorithm and Adaptive double threshold algorithm, each of which has advantage over others in different channel conditions and depending upon many other factors. The problem statement of this Dissertation work is to produce a comparative performance analysis between the many energy based spectrum sensing techniques in different channel conditions. The simulations will be carried out for AWGN channel with constant noise power and variable noise power (noise uncertainty condition).

1.5 Objectives

The objectives of this book are in two folds:

1) Analytically compare the performance of Energy detection based spectrum sensing based on three algorithms namely Double Threshold Energy Detection Algorithm, Adaptive Spectrum Sensing Algorithm, and Adaptive Double Threshold Energy Detection Algorithm.
2) Verify the analytical results through simulation results.

1.6 Organization of the Dissertation

Chapter 2 provides some related work in the literature survey for the problem proposed.

Chapter 3 presents the spectrum sensing methodologies and performance of energy detection based on single threshold algorithm, double threshold algorithm both in cooperative and non-cooperative environment Adaptive Spectrum Sensing when noise is uncertain and adaptive double threshold algorithm in non –cooperative environment analytically.

Chapter 4 depicts the overall system model, simulation steps and the simulation flowchart.

Chapter 5 gives discussion of simulation results of different Energy detector based spectrum sensing methods for non-cooperative Cognitive Radio networks.

Chapter 6 highlights the overall conclusion and future recommendation which can be taken up in the field of spectrum sensing.

2 Literature Review

2.1 Introduction

The recently emerged wireless technologies are Wi-Fi, Wimax, Bluetooth, Zigbee. The advancement in cellular voice and digital services, broadcast satellite is remarkable. Due to large number of services, the spectrum they use for radio waves, the spectrum scarcity problem is prominent. The fixed spectrum allocation policy, however, does not allow unlicensed user to use the licensed spectrum. The assigned spectrums to some of the applications are used inefficiently as far as time and geography is concerned. The secondary user can make use of those bands which are inefficiently used. This is the basic idea of cognitive radio.

2.2 Cognitive radio

(CR) is an emerging wireless communication technology based on software defined radio which is intelligent enough to adapt its operating parameters (i.e. transmitter and receiver parameters) to communicate efficiently without interfering to the licensed users.

2.2.1 Brief History of Cognitive radio

The idea of CR was first proposed by Joseph Mitola III and Gerald Q. Maguire, J in 1999 A. D. through an article. In the article, Mitola described how a cognitive radio could enhance the flexibility of personnel wireless service through a new language called the Radio Knowledge Representation Language (RKRL) [2]. The concept of RKRL was further expanded in his own doctoral dissertation. This language represents knowledge of radio etiquette, devices, software modules, propagation, networks, user needs, and application scenarios in a way that supports automated reasoning about the need of the user. His dissertation presented a conceptual overview of cognitive radio as an exciting multidisciplinary subject.

After the pioneering work of Mitola, the Federal Communications Commission (FCC) published a report prepared by the Spectrum policy Task Force, aiming at improving the way in which the previous resource is managed in the United States [5]. One of the major findings and recommendations of the report is rather revealing in the context of spectrum utilization: "In many bands, spectrum access is a more significant problem than physical scarcity of spectrum, in large part due to legacy command and control regulation that limits the ability of potential spectrum users to obtain such access." Then FCC officially has started to do research on spectrum policy and played vital role in the development of the cognitive radio.

Cognitive radio technology aimed at improving utilization of the radio spectrum is further defined by Haykin as "Cognitive radio is an intelligent wireless communication system that is aware of its surrounding environment (i.e., outside world), and uses the methodology of understanding by building to learn from the environment and adapt its internal states to statistical variations in the incoming RF stimuli by making corresponding changes in certain operating parameters (e.g., transmit power, carrier frequency, and modulation strategy) in real time, with two primary objectives in mind: Highly reliable communications whenever and wherever needed, Efficient utilization of the radio spectrum." Moreover, FCC has presented detailed expositions of signal processing and adaptive procedures that lie at the heart of cognitive radio. In cognitive radio, licensed users (a.k.a. secondary users) "opportunistically" operate in fallow licensed spectrum bands without causing interference to licensed users (a.k.a. primary or incumbent users), thereby increasing the efficiency of spectrum utilization. This method of sharing is often called Opportunistic Spectrum Sharing (OSS). CRs are seen enabling technology for OSS. Unlike a conventional radio, a CR has the capability to sense and understand its environment and proactively change its mode of operation as needed. CRs are able to carry out spectrum sensing for the purpose of identifying fallow licensed spectrum i.e., spectrum "white spaces". Once white spaces are identified, CRs opportunistically utilize these white spaces by operating in them without causing interference to primary users. As the cognitive radios are allowed to make use of the bands of primary users, the problem of spectrum scarcity can be solved by its efficient implementation.

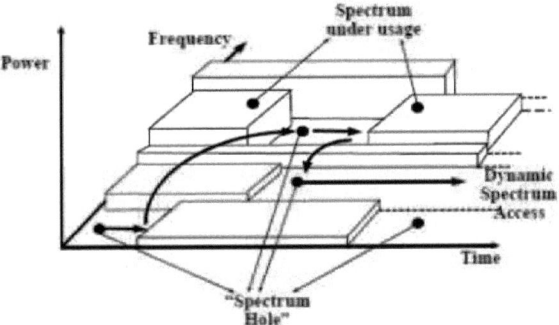

Figure 2.1: Spectrum hole concept [20]

Figure 2.1 shows the concept of white spaces or spectrum holes in which secondary or cognitive users make use for their tasks. It also shows how SUs move to another spectrum hole, when primary resumes for avoiding interference to them [7]. Thus, the task is not only to identify the spectrum hole but also to find out the best holes to decide the usage of the unused spectrums. Many factors such as frequency selection, modulation schemes, and power level needed to be considered to sense the variations in the radio environment so as to avoid interference to other users. CR promises all these functions and helps to utilize the spectrum band efficiently.

2.2.2 Features of Cognitive radio

The Federal Communications Commission (FCC) has identified the features that CRs can incorporate to enable a more efficient and flexible usage of free spectrum. The features are [5]:

Frequency Agility: - It is the ability of the CR to change its operating frequency to optimize its use in adapting to the environment.

Dynamic Frequency Selection (DFS):- The radio is able to detect the signals from the transmitters so as to choose the optimal operating environment.

Adaptive Modulation: - This feature enables the CR to modify the transmission characteristics and waveforms so that it can gain the opportunities for using the spectrum.

Transmit Power Control (TPC):-This feature allows CR to transmit both at full power and at lower power when necessary.

Location Awareness: - The CR is able to identify its location as well as the location of other devices operating in the same spectrum to optimize transmission parameters in order to increase the spectrum re-use.

Negotiated Use: -The CR may have some mechanism to enable the sharing of spectrum under prearranged agreements between a license and a third party or on an ad – hoc or real – time basis.

Thus, the most important components of cognitive radio concept is the ability to measure, sense, learn, and be aware of the parameters related to the radio channel characteristics, availability of spectrum and power, radio's operating environment, user requirements and applications, available networks (infrastructures) and nodes, local policies and other operating restrictions.

2.3 Function of Cognitive Radio

The CR network has two components: the primary network and the secondary one. The primary network, also referred to as licensed network, has a license to operate in a certain frequency band. This consists of primary users (PUs) with or without primary base stations (BSs). PUs are not equipped with any CR functions, whereas, the secondary network is able to share or access the licensed spectrum without affecting the primary

12

network transmission. The secondary network is composed of secondary users (SUs) with or without base station. Additionally, spectrum broker can be used to enable efficient and fair spectrum sharing between multiple secondary networks coexist in the same frequency band. To support this type of spectrum sharing between the primary and cognitive networks, and to guarantee efficient usage of the resources in both networks, CR is required to perform the following four functions [8]. They are:-

  i) Spectrum sensing
  ii) Spectrum decision
  iii) Spectrum sharing
  iv) Spectrum mobility

2.3.1 Spectrum sensing

By this function, the CR monitors its radio environment in order to identify the PUs activity. Based on the sensing information, CR can determine the available spectrum holes that can be used for the CR transmission in a particular time, frequency and location. Furthermore, the CR need to keep sensing the frequency spectrum during the CR transmission to avoid interfering with re-appeared PUs. Spectrum sensing can be performed in either centralized or distributed ways. In centralized spectrum sensing, a central unit, also called sensing controller, is in charge of the sensing process. The sensing information is shared with different SUs using a control channel. Although that the centralized approach reduces complexity of the SUs devices, it suffers from hidden or far PUs detection problem. The SUs are performing the spectrum sensing in the distributed way. Depending on the level of cooperation in the network each SU can take the decision based on his sensing information (non - cooperative sensing) or based on the sensing information shared with others SUs in the network (cooperative sensing). Further, a central unit can collect the distributed sensing information to control the cognitive traffic. The cooperative spectrum sensing is more accurate and can reduce the primary signal detection time. However, cooperative sensing introduces additional signaling overhead which increases with the number of SUs and with fast varying spectrum usage.

2.3.2 Spectrum decision

This function analyzes the information from the spectrum sensing phase. The characteristics of the detected spectrum holes, the probability of the PU appearance, and the possible sensing errors should be considered before making the spectrum access decision. Once the appropriate band is selected, the CR has to optimize the available system resources in order to achieve the required objective.

2.3.3 Spectrum sharing

This function chooses the appropriate MAC protocol to access the spectrum holes. By the MAC protocol, fair spectrum sharing between the different SUs can be guaranteed. Further, coordination between nodes can be achieved in order to avoid the collision with PUs as well as other SUs.

2.3.4 Spectrum mobility

It is also called spectrum handover and by this function, CR is able to change the operating band in the case when primary wanted to use its band. Furthermore, the CR can perform the spectrum handover in order to improve the secondary network performance by transmitting in another spectrum hole with better condition. The protocol parameters at the different levels should be adapted according to the new operating band.

2.4 Physical Architecture of Cognitive Radio

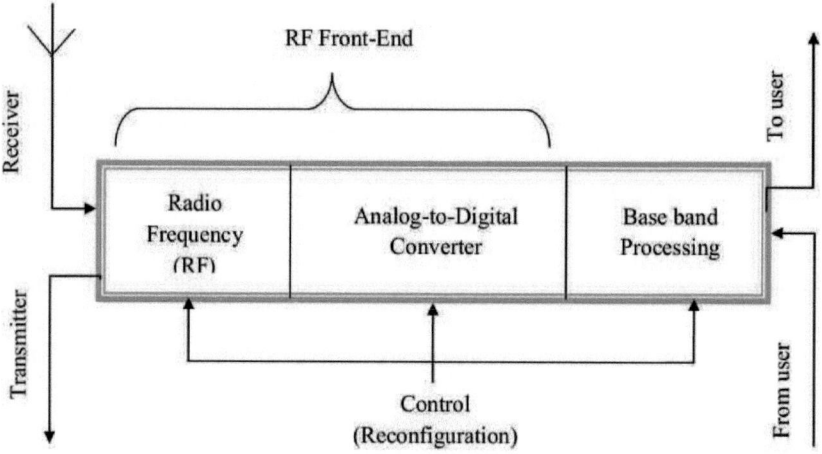

Figure 2.2: Cognitive radio transceiver [20]

Figure 2.2 shows the generic architecture of cognitive radio transceiver. As shown in the figure, the main parts of cognitive radio transceiver are:

i)   Radio front end
ii)   Baseband processing Unit

There is a control unit for adapting the radio frequency (RF) front end so that each component can be reconfigured accordingly. In the RF front-end, the received signal is amplified, mixed, and analog to digital (A/D) converted. In the baseband processing unit, the signal is modulated/demodulated and encoded/decoded. It is more similar to the

14

existing transceivers. The newness of cognitive radio transceiver is its RF front-end. The wideband sensing capability of this front end is its unique feature. This function is mainly related to RF hardware technologies such as wideband antenna, power amplifier, and adaptive filter. RF hardware for CR should be capable of tuning to any part of a large range of frequency spectrum so as to take the real-time measurements of the spectrum information from the radio environment.

The wideband front-end architecture for the CR has the following structure as shown in Figure 2.3:

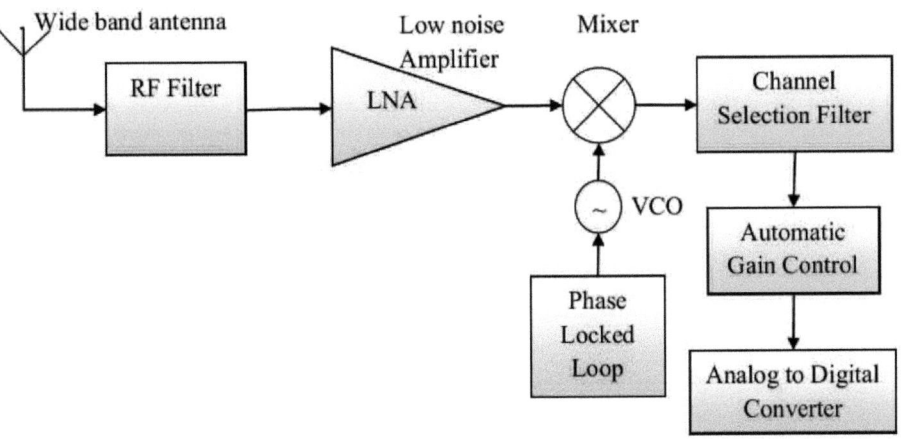

Figure 2.3: Radio front-end [20]

i)   RF filter: - It selects the desired band by band pass filtering of the received RF signal. It is also called adaptive multiband filter.

ii)   Low noise amplifier (LNA):- It amplifies the desired signal at the same time minimizing the noise component.

iii)   Mixer: - In the mixer, the locally generated signal is mixed with the received signal and converted to the baseband or intermediate frequency (IF).

iv)   Voltage controlled oscillator (VCO):- It generates the signal at specific frequency for a given voltage to mix with the incoming signal, thus converting the incoming signal to baseband or intermediate frequency.

v)   Phase locked-loop (PLL):- It ensures that the signal is locked on a specific frequency and can also be used to generate precise frequencies with fine resolution.

15

vi)   Channel selection filter: - This filter is used to select the desired channel and to reject the adjacent channels. There are two types of channel selection filter. The direct conversion receiver uses a low-pass filter for channel selection on the other hand; the super heterodyne receiver adopts a band pass filter.

vii)   Automatic gain control (AGC):- It maintains the gain or output power level of an amplifier constant over a wide range of input signal levels.

2.5 How is a Cognitive Radio Different from other Radios

Traditional radio has hard-wired to communicate using one specific protocol. Traditional hardware based radio devices limit cross-functionality and can only be modified through physical intervention. This results in higher production cost and minimal flexibility. A software defined radio (SDR) is radio communication system where components that have been typically implemented in hardware (e.g. mixers, filters, amplifiers, modulators/demodulators, detectors etc.) are instead implemented by means of software on a personnel computer or embedded system. This makes SDR tremendously versatile. A cognitive radio is an intelligent radio that can be programmed and configured dynamically. Its transceiver is designed to use the best wireless channels in its vicinity. Thus, this has the capability of both reconfigurability and awareness.

2.6 Cognitive Radio Applications, Advantages and Disadvantages

The applications of cognitive radio can be short range wireless communications such as Multimedia Downloads, Emergency communication, Broadband Wireless services, Multimedia Wireless networking etc.

The cognitive radio has many advantages. The advantages are pointed out as follows:

i)   It is expected to be a powerful tool for mitigating and solving spectrum access issues.
ii)   It improves current spectrum utilization.
iii)   It improves performance of wireless network through increased user throughput and system reliability.

The disadvantages are listed as follows:

i)   It adds extra security challenges.
ii)   It adds extra regulatory concerns.
iii)   Fear of undesirable adaptation.

2.7 Cognitive radio standardization

The standardization is a key aspect of the success of the current and future CR systems. IEEE started the development of the first international CR standard. Meanwhile, IEEE has several ongoing works to improve the current standards to support the cognitive capability. In addition to the underway work of IEEE, International Telecommunication Union (ITU), European Telecommunication Standards Institute (ETSI), Defense Advanced Research Projects Agency (DARPA) [4] and European Association for Standardizing Information and Communication systems (ECMA) are examples of other international organizations or associations who have made contributions in making standards for various applications. Within the IEEE, two major standards on CR are IEEE 802.22 and IEEE P1900.

IEEE 802.22: This standard is the first worldwide standard on CR technology. The main target of this standard is to enable the sharing of the TV spectrum with broadcast device in the low population rural areas. In this standard, not only the PHY and MAC layers are considered but also it addresses an additional functions like the spectrum sensing functions and the geo – location one. Using the spectrum sensing function, the spectrum holes are identified while the geo – location one is determining the location of the cognitive devices. The location information is combined with a database of the primary transmitters to determine the available channels. The network BS is covering a geotropically area with 30 km radius and can support a maximum of 255 fixed units of customer premises equipment (CPE). The minimum downlink (BS to CPE) throughput is 1.5 Mb/s while the minimum in the uplink (CPE to BS) is 384 kb/s.

IEEE P1900: This standard focuses on the next generation radio and spectrum management. The standard considers the advanced radio system technologies such as the CR systems, policy defined radio system, adaptive radio systems and related technologies. Moreover, the standards consists of six working groups: IEEE P1900.1 to define the glossary of the terms, IEEE P1900.2 for the interference coexistence analysis, IEEE P1900.3for the evaluation of software modules in SDR to guarantee the compliance in the software part, IEEE P1900.4 is the major working group which relates to coexistence support for the reconfigurable heterogeneous air interference, IEEE P1900.5 for the definition of the policy language and policy architectures, and finally, IEEE P1900.6 to define the spectrum sensing interference as well as data structures for DSA systems.

Furthermore, within the IEEE 802 standard committee, the wireless coexistence technical advisory group IEEE 802.19 is established to deal with the issue of the coexistence of different wireless networks within the same location.

2.8 Scope Area

Out of the four functions of CR, spectrum sensing is found to be an active area of research. In this book, energy based detection is studied. Since energy detection (ED) is easy to implement and requires no prior knowledge about the primary signal, it is the most common type of spectrum sensing.

Decision making criteria in ED is

$$x(t) = \begin{cases} n(t) & : H0 \\ n(t) + hg(t) & : H1, \end{cases} \qquad (2.1)$$

where,$x(t)$ is the received signal at the secondary, $g(t)$ is the signal transmitted from PU, $n(t)$ is the additive white Gaussian noise and h is the channel gain from primary user's transmitter to the secondary user's receiver. $H0$ hypothesis means there is no primary user present in the band, while $H1$ means the primary user's presence. The detection statistics of the energy detector can be defined as the average or total energy of N observed samples

$$T = (1/N) \sum_{t=1}^{N} |x(t)|^2 \qquad (2.2)$$

The decision on whether the spectrum is being occupied by the primary user is made by comparing the detection statistics T with the predefined threshold λ. The performance of the detector is characterized by two probabilities: the probability of false alarm P_F and the probability of detection P_D. P_F denotes the probability that the hypothesis test decides H1 while it is actually H0, i.e.

$$P_F = P_r(T > \lambda|H_0) \qquad (2.3)$$

P_D denotes the probability that the test correctly decides H_1, i.e.,

$$P_D = P_r(T > \lambda|H_1) \qquad (2.4)$$

A good detector should ensure a high detection probability P_D and a low false alarm P_F, or it should optimize the spectrum usage efficiency (e.g., QoS of a secondary user network) while guaranteeing a certain level of primary user protection. There are various approaches being proposed to improve the efficiency of energy detector based spectrum sensing. Since the detection performance is very sensitive to noise power estimate error, an adaptive noise level estimation approach is proposed, where multiple signal classification algorithms is used to decouple the noise and signal sub – spaces and estimate the noise floor [9]. A constant false alarm rate criteria or constant detection criteria are used for calculating the decision threshold. A well-chosen detection threshold can minimize spectrum sensing error, provide the primary user with enough protection, and fully enhance spectrum utilization [10]. The detection threshold is optimized iteratively to satisfy the requirement on false alarm probability [11]. Threshold optimization subject to spectrum sensing constraints is investigated, where an optimal adaptive threshold level is developed by utilizing the spectrum sensing error function [12]. In order to find and localize the narrowband signals, a localization algorithm based on double thresholding (LAD) is proposed, where the usage of two thresholds can provide signal separation and localization [13]. The analysis showed that detection of narrowband transmission using energy detection over multi – band orthogonal frequency – division multiplexing (OFDM) is feasible, and can be further extended to cover more complex system [14].

Besides its low computational and implementation complexity and short detection time, there also exist some challenges in designing a good energy detector. First, the detection threshold depends on noise power, which may change over time and hence is difficult to measure precisely in real time. In low signal – to – noise ratio (SNR) regimes where the noise power is very high, reliable identification of a primary user is even impossible [15]. Moreover, an energy detector can only decide the primary user's presence by comparing the received signal energy with a threshold; thus, it cannot differentiate the primary user from other unknown signal sources. As such, it can trigger false alarm frequently [17]. Thus, not only the robustness issues related to spectrum sensing as a whole are very crucial but also robustness in case of energy detector is equally a weighted issue.

The robustness of energy based detector can be described in two folds: first, the SUs transmission should not be interfering with the transmission of the PUs and second, the legitimate SUs or say good SUs should have to use the fallow spectrum dynamically. Therefore, the first issue is related to how efficiently the primary transmission is detected (related to reliability) and the latter is related to the throughput of the secondary network (related to efficiency) [16].

3 Spectrum Sensing Algorithms

Spectrum sensing is the most basic functionality of cognitive radio system. It is a process of detecting the spectrum holes or white spaces (unused spectrums) and sharing them among cognitive users without causing any harmful interference to both primary network and cognitive network itself. In cognitive radio technology, two users are defined. First one is called primary users which have higher priority or legacy rights on the usage of a specific part of the spectrum. The second is secondary users which have lower priority on the usage of this spectrum. They exploit the spectrum in such a way that they do not cause any interference to the primary users. Spectrum sensing includes awareness about the interference temperature and existence of primary users (PUs).

This work focuses on spectrum sensing. It not only involves obtaining the spectrum usage characteristics across multiple dimensions such as time, space, frequency and code but also involves determining what type of signals are occupying the spectrum (including the modulation waveform, bandwidth, carrier frequency etc.). The present development of spectrum sensing is still in its early stage. A number of different methods are proposed for identifying the presence of signal transmission. In some approaches, characteristics of the transmitted signal are detected for deciding the signal transmission as well as identifying the signal type. The most commonly used spectrum sensing techniques are:

i)   Matched filtering
ii) Cyclostationarity – based sensing
iii) Waveform –based sensing
iv) Energy Detector-based sensing

3.1 Matched filtering

It can be the optimal signal detection, when the secondary user(s) has/have prior knowledge of the PU signal, as it maximizes the Signal -to -Noise Ratio (SNR).Filtering is obtained by correlating a known signal, or template, with an unknown signal to detect the presence of the template in the unknown signal. This equivalent to convolving the unknown signal with a time reversed version of the template. The main advantage of matched filter is that it needs less time to achieve high processing gain due to coherent detection. The significant disadvantage of the matched filter is that it would require a dedicated sensing receiver for all primary user signal types. In the CR scenario, however, the use of matched filter can be severely limited since the information of the PU signal is hardly available at the CRs. It has implementation complexity. It consumes large power as various algorithm need to executed for detection. The use of this approach is still possible if we have partial information of the PU signal such as pilot symbols or preambles, which may be used for coherent detection. For instance, to detect the presence of a digital television ((DTV) signal, we may detect its pilot tone by passing the DTV signal through a delay-and-multiply circuit. If the squared magnitude of the output signal is larger than a threshold, the presence of DTV signal can be detected.

3.2 Cyclostationarity-based Sensing

Here, sensing is done by exploiting the cyclostationarity features of the received signals. Cyclostationary features are caused by the periodicity in the signal or in its statistics like mean and autocorrelation. Since it can differentiate primary users' signals from noise, it is more robust than energy detector when SNR is low or noise is uncertain [24]. The fact that the noise is wide-sense stationary (WSS) with no correlation while modulated signals are cyclostationary with spectral correlation due to the redundancy of signal periodicities make possible to differentiate noise with signal. The detection process can be explained as follows:

First, finding the cyclic spectral density (CSD) based on cyclic auto-correlation function (CAF) as $S(f,\alpha) = \sum_{t=-\infty}^{\infty} R_y^\alpha (t) e^{-j2\pi ft}$ where $R_y^\alpha(t) = E[y(n+t)Y^*(n-t)e^{j2\pi\alpha n}]$ is cyclic auto-correlation function, and α is the cyclic frequency.

3.3 Waveform-Based Sensing

It depends upon the known patterns such as preambles, midambles, regularly transmitted pilot patterns, spreading sequences etc. In the presence of known patterns, sensing can be performed by correlating the received signal with a known copy of itself. It is proved to outperform the energy detector as far as reliability and convergence time is concerned. Its performance can be enhanced with increasing the length of the known signal patterns. It requires short measurement time but it is equally susceptible to synchronization error.

3.4 Energy Detector-based sensing

It is very generic since receivers do not need any prior knowledge of the primary users' signal. In CR scenario, where early information of PU signal is unknown, this approach may be an optimal detection method. In the energy detection approach, the radio frequency (RF) energy in the channel or the received signal strength indicator is measured to determine whether the channel is idle or not. First, the input signal is filtered with a band pass filter to limit the signal within the bandwidth of interest. The output signal is then squared and integrated over the observation time. Lastly, the output of the integrator is compared to a predetermined threshold to decide for and against the PU signal. Although it has low computational and implementation complexities, it still offers some challenges. The first problem is that it has poor performance under low SNR conditions. This is because the noise variance is not accurately known at the low SNR, and the noise uncertainty may render the energy detection useless. Another challenging issue is the inability to differentiate interference from PUs and noise. Furthermore, the selection of decision threshold is challenging since it depends on noise variance and small noise estimation error can result in significant performance loss.

Since it very simple and easy to implement, and more than this it does not require the prior information of the PU signal, it popular in cognitive radio scenario. The main components in the energy detector are depicted in the Figure 3.1.

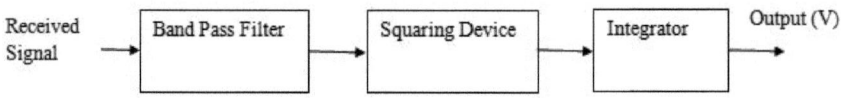

Figure 3.1: Block diagram of energy detector

The main components are:

i)   Band pass filter or pre-filter
ii)   Squaring device
iii)   Integrator

The output of the energy detector i.e. the output of the integrator is the filtered received signal squared and averaged over the time interval 'T'. This output is considered the test statistic to test the previously stated hypotheses H_0 and H_1. The two hypotheses are defined as: H_0 = primary user is absent, H_1 = primary user is present. The spectrum sensing, in fact, is the first step that is needed to perform for communication to make possible. It is simply treated as identification problem, modeled as binary test or hypothesis test. The sensing devices have to just decide between for one of the two hypotheses. Let us suppose x(t), n(t) and y(t) respectively denotes the signal transmitted by primary users, additive white Gaussian noise with variance and signal received by the secondary user(s). Then the two hypotheses are stated as H_1: $y(t) = x(t) + n(t)$ and H_0: $y(t) = n(t)$. Hypothesis H_0 indicates absence of primary user at that frequency band of interest. This band has only noise. However, H_1 points towards presence of primary user. The decision information of either hypothesis maybe true or false resulting in a number of situations with associated probabilities. For example, Probability of taking decision H_1 given, primary user actually exist i.e.,$P(H1/H1)$. This scenario is described by probability of detection (P_d).Probability of taking decision H_0, given, primary user actually exists i.e.,$P(H0/H1)$. This scenario defines the probability of miss detection (Pm).The other can the probability of taking decision H_1, given, primary user actually does not present i.e., $P(H1/H0)$. This scenario defines the probability of false alarm (Pf).These two probabilities are depicted in the Figure 3.2 for clarifications.

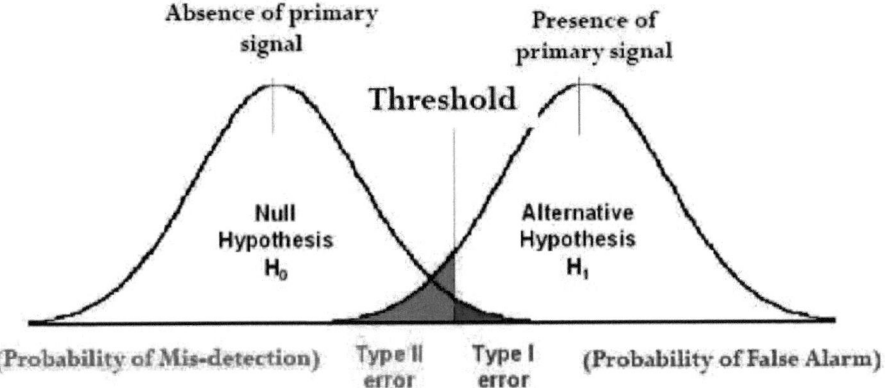

Figure 3.2: Definition of probability of detection and probability of false alarm

The probability of detection is the main concern as it gives the probability of correctly sensing for the presence of PUs in the frequency band. The probability of miss detection is just the complement of probability of detection. The aim of the sensing schemes should be to maximize the probability of detection for small value of probability of false alarm as possible. But there is a trade-off between these probabilities. The performance of the sensing schemes is judge based on the information reflected in the receiver operating characteristics (ROC) curve [21]. This presents the valuable information as regards the behavior of probability of detection with change of probability of false alarm. Also, the curve that shows the behavior of probability of miss detection versus probability of false alarm is called complementary ROC curve. Thus, ROC curves allow exploring the relationship between the sensitivity (probability of detection) and specificity (probability of false alarm) of the energy detection for different system parameters of interest.

3.4.1   Conventional single-threshold energy detection algorithm

In this algorithm, as the name suggest, there is a single detection threshold. When the received signal energy V is greater than the detection threshold V_{th} , the detector concluded that the primary is present in the band of interest, and depicted as H_1. On contrary, the primary user is not presented, and depicted as H_0. The detection probability, false alarm probability and miss probability respectively can be calculated as:

$$P_D = Pr\{V > V_{th}|H_1\} = Q_u\left(\sqrt{2\gamma}, \sqrt{V_{th}}\right) \tag{3.1}$$

$$P_{FA} = Pr\{V > V_{th}|H_0\} = \frac{\Gamma\left(u, {}^{V_{th}}\!/_2\right)}{\Gamma(u)} \tag{3.2}$$

$$P_M = Pr\{V \leq V_{th}|H_1\} \tag{3.3}$$

The detailed derivations are given in Appendix A. γ is the Signal-Noise Ratio (SNR) received by cognitive user, V_{th} is the detection threshold, $Q_u(a,b)$ is normalized Marcum function of order u, a monotonically increasing function with u and monotonically decreasing with b; $\Gamma(a,b)$ is a non-complete gamma function, monotonically decreasing with b and $\Gamma(a)$ is complete gamma function. From (3.2): $P_{FA}.\Gamma(u) = \Gamma\left(u, {V_{th}}/{2}\right)$.Based on the definition of non-complete and complete gamma function (As given in Appendix B):

$$\Gamma\left(u, {V_{th}}/{2}\right) = \int_{V_{th/2}}^{\infty} t^{u-1} * e^{-t}\,dt$$

$$= \int_{V_{th/2}}^{0} t^{u-1} * e^{-t}\,dt + \int_{0}^{\infty} t^{u-1}.e^{-t}dt$$

$$= \Gamma(u) - \int_{0}^{V_{th/2}} t^{u-1} * e^{-t}\,dt$$

Again $\int_{0}^{V_{th/2}} t^{u-1} * e^{-t}\,dt = \Gamma(u)p\left(u, V_{th/2}\right)$ where $z = p(u,x) = \frac{1}{\Gamma(u)}$ is low order non-complete gamma function, and monotone decreasing with x, moreover, $x = p^{-1}(u,z)$ is monotone increasing function with z [22].

So, $P_{FA}.\Gamma(u) = \Gamma(u) - \Gamma(u).p\left(u, V_{th/2}\right)$

$$or, P_{FA} = 1 - \Gamma(u).p\left(u, V_{th/2}\right)$$

$$or, V_{th} = 2p^{-1}(u, 1 - P_{FA})$$

On substituting V_{th} into equation (3.1) and (3.3), we get,

$$P_D = Q_u(\sqrt{2\gamma}, \sqrt{V_{th}}) = Q_u\left(\sqrt{2\gamma}, \sqrt{(2p^{-1}(u, 1 - P_{FA}))}\right) \tag{3.4}$$

$$P_M = 1 - P_D = 1 - Q_u\left(\sqrt{2\gamma}, \sqrt{(2p^{-1}(u, 1 - P_{FA}))}\right) \tag{3.5}$$

From above analysis, the single-threshold energy detection algorithm may cause serious interference to the primary user. Probability of detection depends on the probability of false alarm and SNR. If SNR is small, detection probability decreases. Also, if the detection probability increase and at the same rate the false alarm probability also increases at a given good value of SNR. The increase in false alarm probability means

24

more interference to the primary [23]. In order to alleviate them, a double - threshold energy detection algorithm has been proposed.

3.4.2 Double threshold energy detector

In this algorithm, another detection threshold is added within the conventional single-threshold energy detection algorithm. The two thresholds are defined as V_{th0} and V_{th1}. The primary user will be detected if and only if $V > V_{th1}$, and will not be present if and only if $V < V_{th0}$. The decisions correspond to H_1 and H_0 respectively. There is high possibility of taking decision if and only if V lies in anywhere in between the thresholds i.e. V is in $(V_{th0}, V_{th1}]$. For the better performance of the detector, it needs re-detection. Based on the conventional single-threshold energy detection algorithm, we can calculate the performance indicator of the double threshold energy detection algorithm such as the detection probability, false alarm probability and missing probability [23]. They can be calculated as:

$$P'_D = P_r(V' > V_{th1}|H_1) = Q'_u\left(\sqrt{2\gamma}, \sqrt{V_{th1}}\right) \tag{3.6}$$

$$P'_{FA} = P_r(V' > V_{th1}|H_0) = \frac{\Gamma\left(u, V_{th1}/2\right)}{\Gamma(u')} \tag{3.7}$$

$$P'_M = P_r(V' \leq V_{th1}|H_1) = 1 - P_D \tag{3.8}$$

where P'_D is the correct detection probability when the primary user is present. P'_{FA} is the probability of the primary user detected presently, given, it is not present. P'_M is the probability of primary user perhaps may not be detected, given, it is present. In double threshold energy detection algorithm, two more performance indicators have been introduced for analysis. They are probability of collision between the cognitive user and the primary user, and the probability of spectrum unavailable to the cognitive user. These two parameters are defined and calculated as follows: The probability of collision between the cognitive user and the primary user: $p_c = p\{V' < V_{th0}|H_1\}$. It is the probability of the primary user which is not detected, but in fact it is existed, and this unoccupied spectrum will be allocated to the cognitive user. It indicates the interference of the cognitive user to the primary user because of the uncertainty of the spectrum detection. The larger the probability of collision between the primary users and the cognitive user is, the more serious the interference of cognitive user to the primary user is, on the contrary, there is less interference. The probability of restricting the cognitive user to the spectrum, that is, the spectrum unavailable probability: $p_{na} = p\{V' > V_{th0}|H_0\}$. It is the probability of primary user may be detected, while in fact it is not present, and this "busy" spectrum should not be allocated to the cognitive user in order to avoid interferences to the primary user. It indicates the efficiency of the spectrum usage,

25

that is, whether there are enough spectrums for the cognitive user to access the system timely. The larger the spectrum unavailable probability is, the less efficiency of the spectrum usage is. On contrary, the spectrum is allocated efficiently. Generally, it has $V_{th0} < V_{th} < V_{th1}$ because of adding detection threshold. We can have form equations (3.1) and (3.6) that $Q'_u\left(\sqrt{2\gamma'}, \sqrt{V_{th1}}\right) < Q_u\left(\sqrt{2\gamma}, \sqrt{V_{th}}\right)$ i.e. $P'_D < P_D$. Similarly we can get: $P'_M > P_M$ and $P'_{FA} < P_{FA}$. It is clear that in the conventional single-threshold energy detection algorithm the probability of collision between the cognitive user and the primary user is also the miss probability, $P_{c1} = P_M$, and the spectrum unavailable probability is also the false alarm probability, $P_{na1} = P_{FA}$. From this, the calculation of probability of collision between the cognitive user and the primary user and the spectrum unavailable probability is done as follows:

$$p_{c2} = P_r\{V' < V_{th0}|H_1\} \quad = 1 - P_r\{V' > V_{th0}|H_1\} = 1 - Q_{u\prime}\left(\sqrt{2\gamma'}, \sqrt{V_{th0}}\right) \quad (3.9)$$

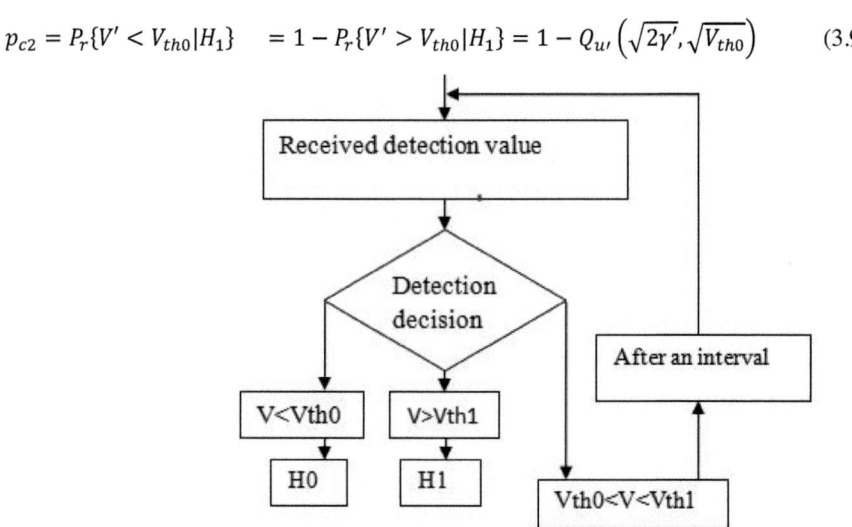

Figure 3.3: Flowchart of double threshold energy detection algorithm

This is the expression to calculate the collision probability. The next is the spectrum unavailable probability as:

$$p_{na2} = P_r\{V' > V_{th0}|H_0\} \quad = \frac{\Gamma(u', V_{th0}/2)}{\Gamma(u')} \quad (3.10)$$

On comparing equations (3.3) and (3.9), we get, $V_{th0} < V_{th}$ and this implies

$$Q_u\left(\sqrt{2\gamma}, \sqrt{V_{th}}\right) < Q_{u\prime}\left(\sqrt{2\gamma'}, \sqrt{V_{th0}}\right)$$

And $p_{c2} < p_{c1}$. Similarly, we can get $p_{n1} < p_{n2}$. It can be concluded that the probability of collision between the cognitive user and the primary user can be decreased effectively, avoiding the cognitive user interfering the primary user. At the same time, this algorithm decreases a little bit spectrum used efficiency. In other words, there is still a possibility to improve the spectrums' used efficiency. The detection algorithm is best described by Figure 3.3.

3.5 Cooperative spectrum sensing

The critical challenging issue in spectrum sensing is the hidden terminal problem, which occurs when the cognitive radio (CR) is shadowed or in severe multipath fading. In this, the CR cannot always sense the presence of PU, and thus it is allowed to access the channel while the PU is still in operation. To solve this problem, multiple CRs can be designed to collaborate in spectrum sensing. Recent research work has shown that cooperative spectrum sensing can greatly increase the probability of detection in fading channels. In general, the cooperative spectrum sensing can be performed as described below: Every CR performs its own local spectrum sensing measurements independently and then makes a binary decision on whether the PU is present or not. All the CRs forward their decisions to a common receiver. The common receiver fuses these decisions and makes final decisions to infer the absence or presence of the PU.

3.5.1 Conventional cooperative spectrum sensing

As we know that the prime goal of spectrum sensing is to distinguish between the two hypotheses. In many papers, it is often assume that there are N secondary users and a fusion center in cognitive radio networks, each secondary user experiences independent and identically distributed fading and shadowing with the same average SNR, and each user has the same threshold value λ. The fusion center receives the information of each secondary user and makes a final decision whether the primary is presence or not. In conventional fusion method, OR-rule is used [25]. For example, if one secondary user observes the primary user, then the fusion center determines that it really exists. Probabilities of detection, missing and false alarm for this cooperative sensing method are as follows,

$$Q_d = 1 - \prod_{i=1}^{N}(1 - P_{d,i}) \qquad (3.11)$$

$$Q_m = \prod_{i=1}^{N} P_{m,i} \qquad (3.12)$$

$$Q_f = 1 - \prod_{i=1}^{N}(1 - P_{f,i}) \qquad (3.13)$$

The individual probabilities are calculated as given by the relations in equations (3.1), (3.2), and (3.3) respectively. Also, Q_d, Q_m and Q_f denote the cooperative probabilities of detection, missing and false alarm respectively, and $P_{d,i}$, $P_{m,i}$, and $P_{f,i}$ are the detection probability, missing probability and false alarm probability of the ith secondary user respectively, and each has the same formulas as described above.

3.5.2 Double threshold energy detection of cooperative spectrum sensing

In conventional energy detections, each secondary user makes their local decisions by comparing its observational value with a predefined threshold as illustrated in Figure 3-4 (a).

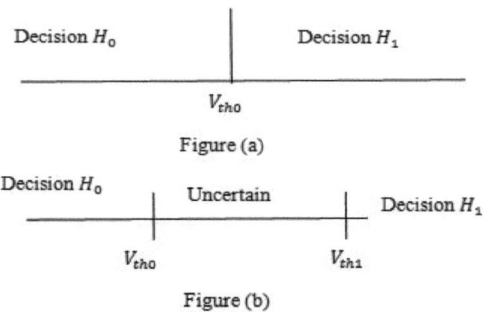

Figure 3.4: Energy detection decision

Here in double threshold, "Region of Uncertainty" represented as $O_i = (V_{th0}, V_{th1})$ indicates the collected energy value of the ith secondary user. Decision H_0 and H_1 will be made when O_i is greater or less than the threshold value V_{th}, respectively. Also it introduced two threshold defined by V_{th0} and V_{th1}. If the energy value exceeds V_{th1}, then this user reports H_1, which means that it sees the primary user. If energy value is less than V_{th0}, decision H_0 will be made. Otherwise, if O_i, is between V_{th0} and V_{th1}, then we also allow the secondary user reporting its observational energy value. So in this model, the fusion center receives two kinds of information: local decisions and observational values of the secondary users, that is, local energy values. Following are the performing schemes of the double threshold energy detection cooperative spectrum sensing method: Each secondary user performs spectrum sensing individually, i.e., energy detection with a result of O_i. If O_i satisfies $V_{th0} < O_i < V_{th1}$, then the ith secondary user sends the energy detection value O_i to the fusion center. Otherwise, it will report its local decision L_i according to O_i[25]. We use R_i to denote the information that the fusion center receives from the ith secondary user, then it can be given by

$$R_i = \begin{cases} O_i & V_{th0} < O_i < V_{th1} \\ L_i & otherwise \end{cases}$$

And
$$L_i = \begin{cases} 0 & 0 \leq O_i \leq V_{th0} \\ 1 & O_i > V_{th1} \end{cases} \tag{3.14}$$

28

Here, we assume that the fusion center receives K local decisions and N-K energy detection values among N secondary users. Then the fusion center makes an upper decision according to N-K energy detection values, which is given by

$$
D = \begin{cases} 0 & 0 \leq \sum_{i=1}^{N-K} O_i \leq V_{th} \\ 1 & \sum_{i=1}^{N-K} O_i > V_{th} \end{cases} \tag{3.15}
$$

where V_{th} is the energy detection threshold value at the fusion center. It shows that these N-K secondary users could not distinguish between the absence and the presence of the primary user, so the fusion center collects their observational values and makes an upper decision instead of the local decision of them [27]. The probabilities of misdetection, false alarm and detection are calculated by the following relations:

$$
Q_m = \sum_{k=0}^{N-1} \binom{N}{K} \prod_{i=1}^{K} P_{m,j} \prod_{i=K+1}^{N} \Delta_{1,i} \left[1 - Q_{(N-K)u}(\sqrt{2\gamma}, \sqrt{V_{th}}) \right] + \prod_{i=1}^{N} P_{m,i} \tag{3.16}
$$

$$
Q_f = 1 - \prod_{i=1}^{N} (1 - \Delta_{0,i} - P_{f,i})
$$

$$
= - \sum_{k=0}^{N-1} \binom{N}{K} \prod_{i=1}^{K} (1 - \Delta_{0,i} - P_{f,i}) \prod_{i=K+1}^{N} \Delta_{0,i} \left[1 - \frac{\Gamma\left[(N-K)u, \frac{V_{th}}{2} \right]}{\Gamma[(N-K)u]} \right]
$$

$$
\tag{3.17}
$$

$$
Q_d = 1 - Q_m \tag{3.18}
$$

Based on the above relations simulations were carried out. The simulation result proved that the performance of double threshold outwit that of the conventional cooperative spectrum sensing methods. But, the detection performance gain was achieved by the increase of communication burdens introduced by the local energy values.

3.6 Adaptive Single-Threshold Energy detection Algorithm

For the large number of samples, the probability is detection and the probability of false alarm are given by $P_D = P_r\{D(Y) > \lambda | H_1\}$ and $P_{FA} = P_r\{D(Y) > \lambda | H_0\}$

For the energy decision threshold (λ), an adaptive decision threshold was set [27]. It is decided by the noise power and signal power, which is to adapt noise fluctuation. The probability of detection and the probability of false alarm are given by

$$P_{d1} = Q\left(\frac{\lambda-(\sigma_n{}^2+\sigma_s{}^2)}{(\sigma_n{}^2+\sigma_s{}^2)\Big/\sqrt{N/2}}\right) \tag{3.19}$$

$$P_{f1} = Q\left(\frac{\lambda-\sigma_n{}^2}{\sigma_n{}^2\Big/\sqrt{N/2}}\right) \tag{3.20}$$

Where $\sigma_n{}^2$ is the noise power and $\sigma_s{}^2$ is the signal power. The optimal threshold λ is decided by the noise power and signal power, which is to adapt noise fluctuation. The smaller λ is, the higher P_{d1} and P_{f1} are. The higher P_{d1} means less interference to PU, but the higher P_{f1} means the less chances of the channel can be reused when it is available and usable. As a result, the achievable throughput for the secondary network is lower. Therefore, there is a tradeoff between P_{d1} and P_{f1}. How to set the decision threshold λ in a robust manner to signal and noise power variation is the key issue of spectrum sensing. If the λ is smaller than the noise power $\sigma_n{}^2$, many samples of the noise would be detected as PU signal, then the probability of false alarm would be very high. Also, if the decision threshold λ is larger than the signal power $\sigma_s{}^2$, many samples of the PU signal would be miss detected, and the probability of detection would be very low. Therefore, we would constrain the decision threshold λ as

$$\sigma_n{}^2 \le \lambda \le \sigma_s{}^2 \tag{3.21}$$

Here the weighted tradeoff principle is used for the well-known two main parameters associated with spectrums sensing performance, i.e., tradeoff between probability of detection and the probability of false alarm. Usually, the design of spectrum detector follows the constant false alarm rate (CFAR) criteria or constant detection rate (CDR) criteria. Generally, it is difficult to achieve the optimal tradeoff between these two probabilities when CFAR principle or CDR principle is applied. This necessitates the weighted tradeoff principle. Let α denote the weight factor of P_{f1} and $1-\alpha$ denote the weight factor of P_{d1}. The weighted probability of miss detection P_{m1} is defined as

$$P_{m1}(\lambda) = \alpha P_{f1} + (1-\alpha)P_{d1} = \alpha Q\left(\frac{\lambda-\sigma_n{}^2}{\sigma_n{}^2\Big/\sqrt{N/2}}\right) + (1-\alpha)Q\left(\frac{\lambda-(\sigma_n{}^2+\sigma_s{}^2)}{(\sigma_n{}^2+\sigma_s{}^2)\Big/\sqrt{N/2}}\right) \tag{3.22}$$

If the bracketed terms are replaced by x_1 and x_2 respectively, then we have

30

$$P_{m1}(\lambda) = \alpha Q(x_1) + (1 - \alpha)Q(x_2)$$

$$= \alpha . \frac{1}{2} erfc\left(\frac{x_1}{\sqrt{2}}\right) + (1 - \alpha)\left[1 - \frac{1}{2} erfc\left(\frac{x_2}{\sqrt{2}}\right)\right]$$

$$= \frac{\alpha}{\sqrt{\pi}} \int_{\frac{x_1}{\sqrt{2}}}^{\infty} e^{-z^2} dz - \frac{1-\alpha}{\sqrt{\pi}} \int_{\frac{x_2}{\sqrt{2}}}^{\infty} e^{-z^2} dz + (1 - \alpha) \tag{3.23}$$

For a given value of α, the weighted probability miss detection is strictly a convex function of λ, so the optimal threshold is given by

$$\lambda^* = \frac{1 + \sqrt{1 + \frac{4(2\sigma_n^2 + \sigma_s^2)}{N\sigma_s^2} ln\left[\frac{\alpha(\sigma_n^2 + \sigma_s^2)}{(1-\alpha)\sigma_n^2}\right]}}{\frac{2\sigma_n^2 + \sigma_s^2}{\sigma_n^2(\sigma_n^2 + \sigma_s^2)}} \tag{3.24}$$

If we know the knowledge of noise power and the signal power or the SNR of the received signal, we can easily get the optimal decision threshold. With this threshold, we can make a good detection decision so that the best performance is obtained.

3.7   Adaptive Double-threshold Energy detection Algorithm

Thus far we have discussed the double threshold energy detection algorithm and adaptive single threshold algorithm.

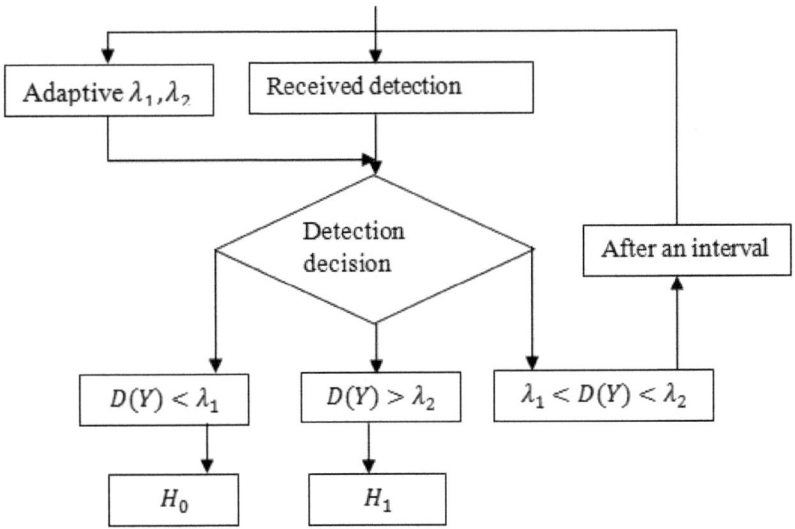

Figure 3.5: Flow chart of adaptive double threshold detection algorithm

Based on these two algorithms, we come up with the double-threshold adaptive spectrum sensing algorithm. As the name suggest, this algorithm has two thresholds (λ_1, λ_2) and they are based on the adaptive threshold which is obtained by estimating the noise power and the signal power. The energy detection process is shown in Figure 3.5. The PU will be detected if the received signal power denoted as $D(Y)$ is greater than the upper threshold denoted by λ_2. It will not be detected if $D(Y) < \lambda_1$, where λ_1 is lower threshold. When the received signal power is in $(\lambda_1, \lambda_2]$, it is prone to mistaken. It needs detection again. This region is called UNCERTAIN REGION [28]. The optimal detection threshold is obtained as that before as given in the Equation (3.21) by the relation:

$$\lambda' = \frac{1 + \sqrt{1 + \frac{4(2\sigma_n^2 + \sigma_s^2)}{N\sigma_s^2} ln\left[\frac{\alpha(\sigma_n^2 + \sigma_s^2)}{(1-\alpha)\sigma_n^2}\right]}}{\frac{2\sigma_n^2 + \sigma_s^2}{\sigma_n^2(\sigma_n^2 + \sigma_s^2)}} \tag{3.25}$$

Now, the issue is to choice the double threshold. We can auto set double threshold λ_1 and λ_2 according to the noise fluctuation based on the optimal threshold. Generally, the double threshold λ_1 and λ_2 are set by

$$\lambda_1 = \alpha\lambda' \tag{3.26}$$

$$\lambda_2 = \beta\lambda' \tag{3.27}$$

where α and β are given constant. We suppose the received power $D(Y)$, and the SNR is γ, then the probability of detection P_D and the probability of false alarm P_{FA} are given by

$$P_D = P\{D(Y) > \lambda_2 | H_1\} = Q_u\left(\sqrt{2\gamma}, \sqrt{\lambda_2}\right) \tag{3.28}$$

$$P_{FA} = P\{D(Y) > \lambda_2 | H_0\} = \frac{\Gamma\left(u, \lambda_2/2\right)}{\Gamma(u)} \tag{3.29}$$

Where SNR $= \frac{\sigma_s^2}{\sigma_n^2}$. We use Δ_0 and Δ_1 to represent the probability of $\lambda_1 \leq D(Y) \leq \lambda_2$ for every cognitive user under hypothesis H_0 and H_1 respectively. Thus, Δ_0 and Δ_1 are given by

$$\Delta_0 = P\{\lambda_1 \leq D(Y) \leq \lambda_2 | H_0\} = \frac{\Gamma\left(u, \lambda_1/2\right)}{\Gamma(u)} - \frac{\Gamma\left(u, \lambda_2/2\right)}{\Gamma(u)} \tag{3.30}$$

$$\Delta_1 = P\{\lambda_1 \leq D(Y) \leq \lambda_2 | H_1\} = Q_u\left(\sqrt{2\gamma}, \sqrt{\lambda_1}\right) - Q_u\left(\sqrt{2\gamma}, \sqrt{\lambda_2}\right) \tag{3.31}$$

And probability of miss detection P_M is given by

$$P_M = P\{D(Y) < \lambda_1 | H_1\} = 1 - \Delta_1 - P_D = 1 - Q_u\left(\sqrt{2\gamma}, \sqrt{\lambda_1}\right) \tag{3.32}$$

The performance analysis of this algorithm can be described as: since the lower value of the double thresholds λ_1 is lower than the single threshold λ', the probability of miss detection P_M would be lower then P'_M as givenby Equation (3.29). On the other hand, the probability of detection can be improved in Equation (3.25). From these two facts we can make a conclusion that the probability of collision can be decreased between the PU and the cognitive user, which can improve the spectrum utilization efficiency.

4 Methodology

4.1 System Description

The system model is developed based on the three algorithm discussed in previous chapter. The system model is shown in the Figure 4.1.

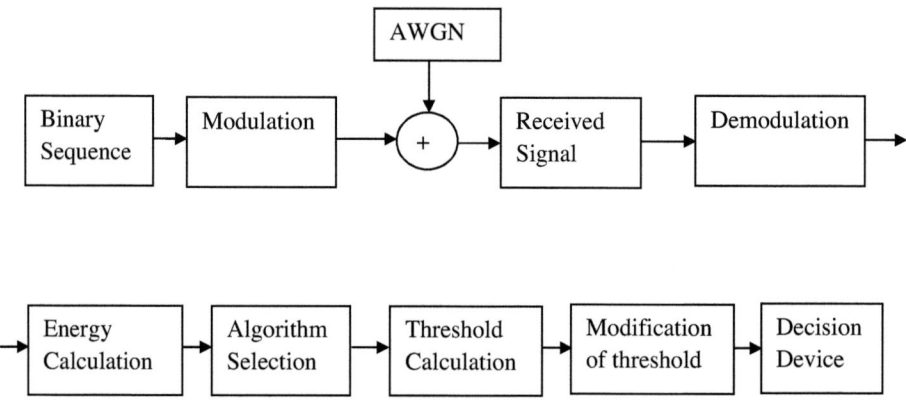

Figure 4.1: System Model

The system model is considered the simple digital communication with QPSK modulation. This QPSK modulated signal is considered as primary signal which is then transmitted from the AWGN channel and received at the secondary receiver after QPSK demodulation. The received signal is then fed into the energy detector which implemented the three different algorithms described earlier. The energy of the received signal is calculated after filtering and averaged over the suitably chosen sample size. The algorithms and their threshold are calculated as per their relationships. Certain modification is done on the calculated threshold for getting its optimal version depending upon the algorithms chosen. At last, the decision device decides for and against the primary user.

In order to verify the performance predicted by the analytical framework discussed in the previous chapter, different parameters such as probability of detection, probability of false alarm, collision probability and spectrum unavailable probability at different conditions are obtained and verified in Matlab. In particular, each calculation and verification is carried out according to the following steps:

1. Decision thresholds (λ_1, λ_2) are generated with respect to constant false alarm rate criteria in case of double threshold algorithm. The optimum threshold is generated in the case when the noise is uncertain (particularly in low SNR) for implementing adaptive spectrum sensing algorithm as given in the Equation (3.24). The two thresholds (λ_1, λ_2) are generated based on optimum threshold as given by the relation Equations (3.26) and (3.27) respectively in implementing adaptive double threshold algorithm.
2. Equally likely hypothesis $H \in \{H_0, H_1\}$ is generated.
3. The received signal from the primary transmitter $y(t) = s(t) + n(t)$ is generated under Additive White Gaussian (AWGN) channel.
4. Next, the received energy i.e. square of $y(t)$ of step 3, at the CR receiver is compared with the respective threshold voltage and respective hypothesis.
5. Steps 1 to 4 are repeated a large number of times (particularly 10000 and above) to reliably estimate the results.

In the later part, we consider SNR and simulate the model as follows

6. Steps 1 to 2 are repeated.
7. CR sensor SNR is generated.
8. The received signal of CR receiver is generated.
9. Step 4 is repeated.
10. Steps 5 to 8 are repeated a large number of times to reliably estimate the probability of detection (P_D), probability of false alarm (P_{FA}), spectrum unavailable probability (P_{na}) and collision probability (P_c)
11. The plot of these parameters versus SNR is generated.

Table 4-1: Simulation parameters used

S.No.	Simulation Parameters	Values
1.	Sample Size (N)	10000 or above
3.	Channel Used	AWGN
4.	Modulation Type	QPSK with modulation index of 4
5.	Number of primary transmitter	1
6.	Signal type	Additive White Gaussian Signal with unity power
7.	Noise type	Additive White Gaussian Noise with random noise power
8.	Constants	$\alpha = 0.8$ and $\beta = 1.2$

With these parameters the simulation is done. The flowchart of the simulation model is shown in the Figure 4.1.

4.2 Simulation flowchart

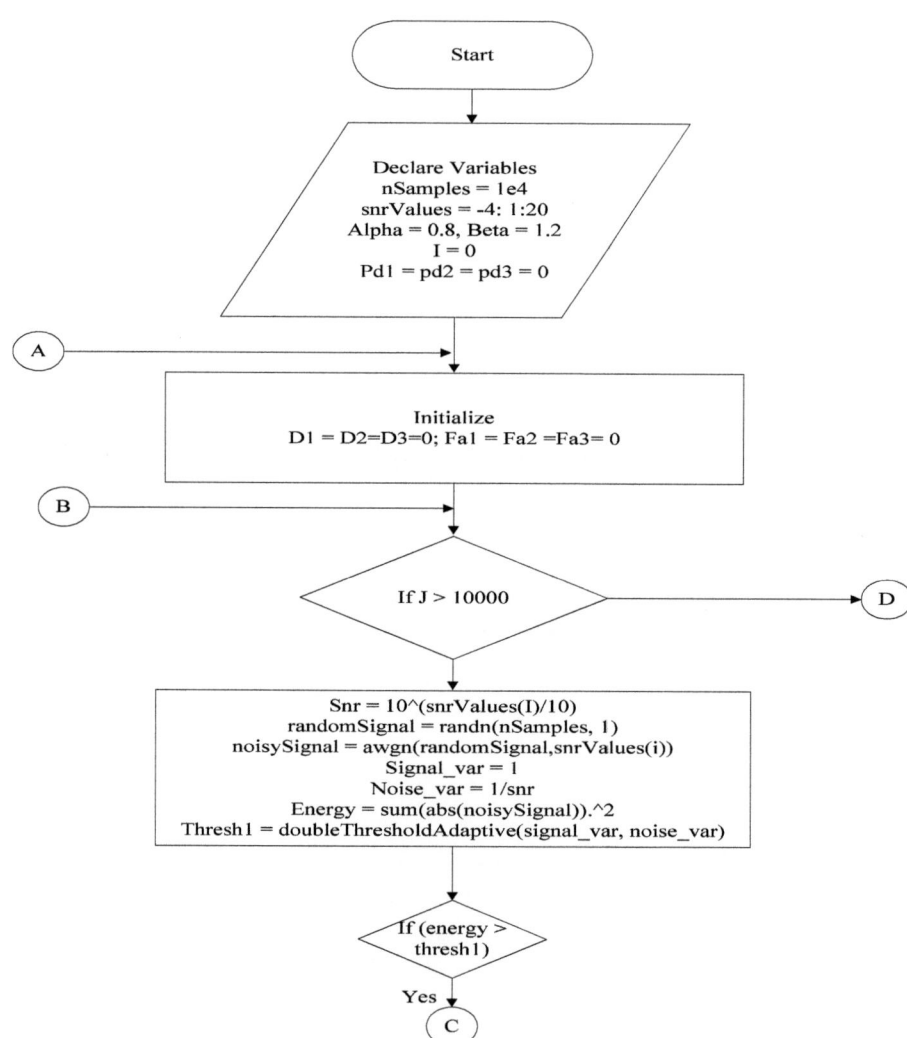

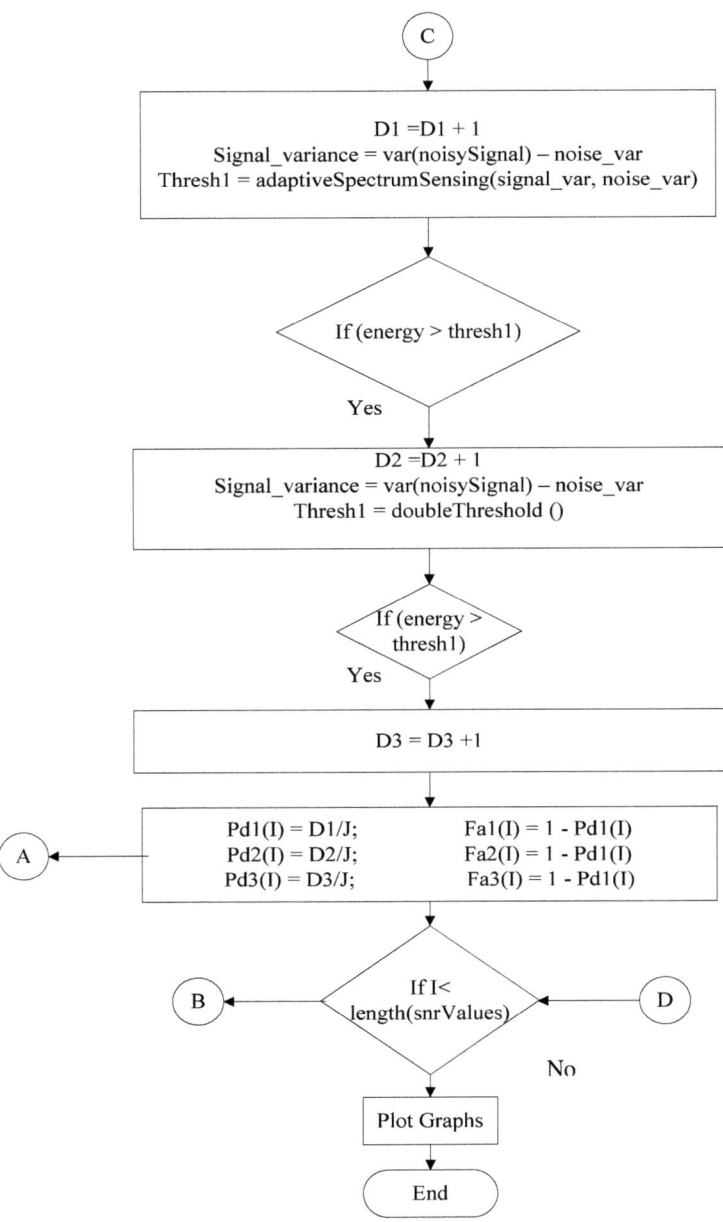

Figure 4.2: Simulation Model

5 Simulation Results and Discussion

5.1 ROC plot for Energy Detector based Spectrum sensing

Let us consider the important parameters and their abbreviation as follows

i) P_M = Probability of missed detection
ii) P_D = Probability of detection
iii) P_{FA} = Probability of false alarm
iv) P_C = Probability of collision
v) P_{na} = Probability of spectrum unavailability

Probability of detection, probability of false alarm and probability of missed detection are the key measurement metrics that are used to analyze the performance of spectrum sensing techniques. The performance of spectrum sensing technique is illustrated by the receiver operating characteristics (ROC) curve which is a plot of probability of detection versus probability of false alarm or probability of missed detection versus probability of false alarm. The latter is also called complementary ROC curve. All the simulations are done using Matlab.

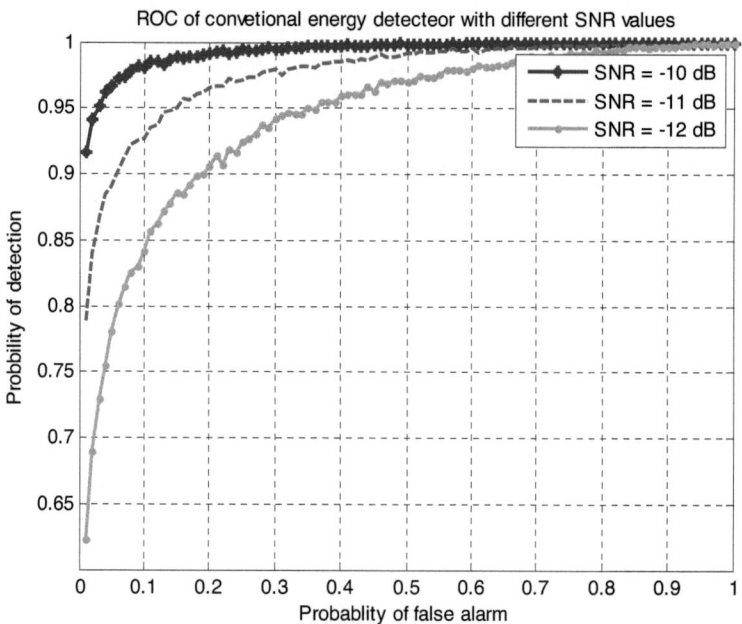

Figure 5.1: ROC curves plot for different values of SNR

5.1.1 Simulation of single threshold energy detection algorithm

The channel used is AWGN. Real valued Gaussian PU's signal is transmitted through AWGN channel. The simulation is done for 10000 times. The sample size taken is 2000 and the SNR (dB) is -12, -11, and -10. The detection performance can be observed by varying the probability of false alarm from 0, 0.1, 0.2,......, 1 and finding the probability of detection by using Monte-Carlo simulation for each case as shown in Figure 5.1: This work describe the receiver through the ROC and complementary ROC curves for different values of probability of false alarm, probability of detection and signal to noise ratio. The effect of sample size is also analyzed.

It is observed that detection performance can be improved by increasing SNR value. This also illustrates the ROC (Receiver Operating Characteristics) curves i.e. P_D versus P_{FA} using energy detection method for spectrum sensing. The graph is plotted for different SNR values over AWGN channel and it also shows that with increase in SNR (Signal to Noise Ratio), the probability of detection increases. This increment is quantified in Table 5-1. Figure 5.2 shows the plot of detection probability for different values of SNR. It shows the detection probability is low in the lower values of SNR and its value goes on increasing with increasing values of SNR. This concludes that the energy detector performance is better only for larger values of SNR.

Table 5-1: Improvement in probability of detection when SNR is increase in Energy detector

Probability of false alarm (PFA)	Probability of detection (PD) (SNR = -12dB)	Probability of detection (PD) (SNR = -11dB)	Improvement (in times)
0.01	0.6358	0.7850	0.23
0.1	0.8427	0.9304	0.10
0.2	0.9066	0.9661	0.065
0.3	0.9379	0.9820	0.047
0.5	0.9724	0.9893	0.0173

From Table 5.1, it is found that with increment of 1 dB in Signal to Noise Ratio, the increment in probability of detection at (SNR = -12dB) is upto 0.23 times as compared to the probability of detection (at SNR = -11 dB) for AWGN channel. So, the conventional energy based detection has low performance for low SNR values.

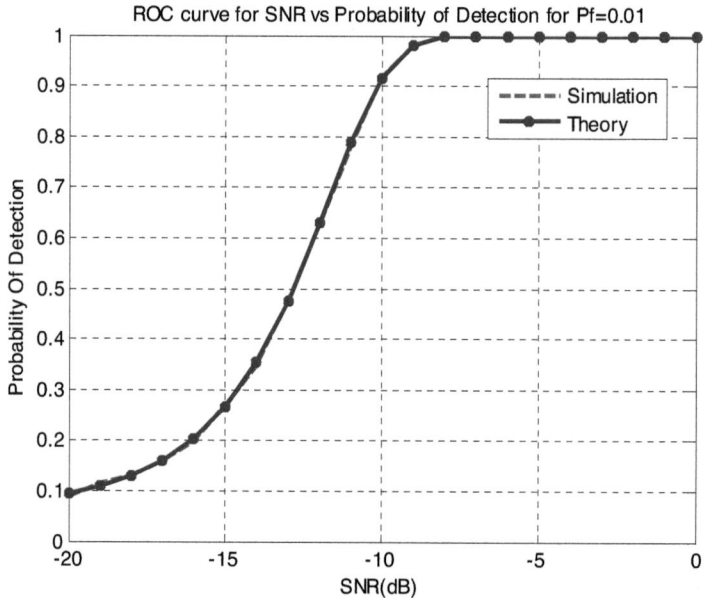

Figure 5.2: Plot of detection probability for different values of SNR

For given value of probability of false alarm, the performance of this detector can be enhanced by increasing the number of sample points even for low values of SNR. This is illustrated in the Table 5-2.

Table 5-2: Improvement of probability of detection for low value of SNR in energy based detection by increasing sample size

Number of samples (N)	Probability of detection (PD) (SNR = -14dB)	Probability of detection (PD) (SNR = -10dB)	Improvement (in times)
600	0.1747	0.5214	1.89
700	0.1901	0.5792	2.04
900	0.2149	0.6570	2.05
1200	0.2523	0.7667	2.03
2000	0.3508	0.9184	1.61

5.1.2 Simulation of Double Threshold energy detection algorithm, Adaptive Spectrum sensing algorithm and Adaptive Double Threshold energy detection algorithm

Algorithm for adaptive double threshold is mentioned as follows: Threshold for the algorithm, as found earlier, is calculated using formula as given by the Equation (3.22).

Then
Threshold1 = α * threshold;
Threshold2 = β * threshold;
If energy > threshold2
        Spectrum Occupied
And if energy < threshold1
Spectrum Free.

The value of α and β are chosen to be 0.8 and 1.2. The noise variance is assumed to be unity to comprise to the double threshold energy detection when the noise is certain. Thus in this case the SNR is 10dB. The signal variance for all case is assumed unity. The modulation scheme used is QPSK. This modulated signal is then passed through the AWGN channel before reaching to the receiver. The sample size taken is 10000. The simulation is done for 20000 times or above for better result.

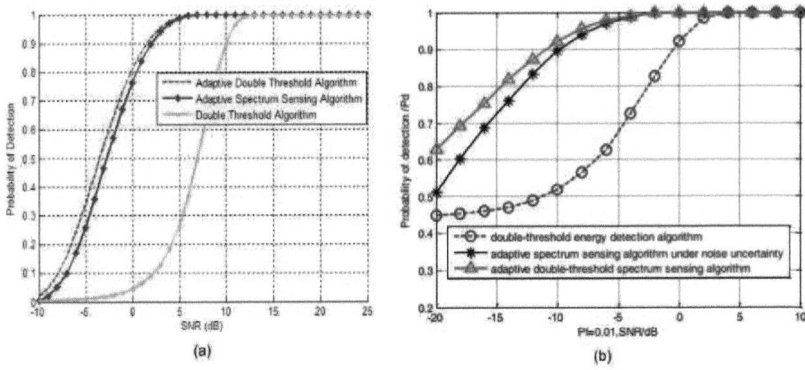

(a)   (b)

Figure 5.3: Plot of probability of detection for different SNR value

Figure 5.3(b) is the output of the base paper [28]. It shows the probability of detection for different values of SNR. From the figure it is clear that Adaptive Double Threshold Spectrum Sensing Algorithm is better than other two algorithms. It also shows that the detection probability for low SNR values is remarkably increased as compared to that in double threshold energy detection algorithm. The performance of the adaptive double threshold algorithm is found better for both the case of low and high values of SNR. The

41

detection probability at low SNR values is high means the interference caused by the secondary to the primary user is significantly lower. Therefore, the interference caused by the secondary to the primary user can be lowered by using the adaptive double threshold algorithm in energy based spectrum sensing. The figure 5.3(a) is the curves obtained from the simulation. Since the parameters used in the simulation are different , more specifically, the channel condition and type, the probability values for given SNR values are different. But, the pattern is the same. That is the tendency of probability of detection is in increasing order with the increment in SNR values.

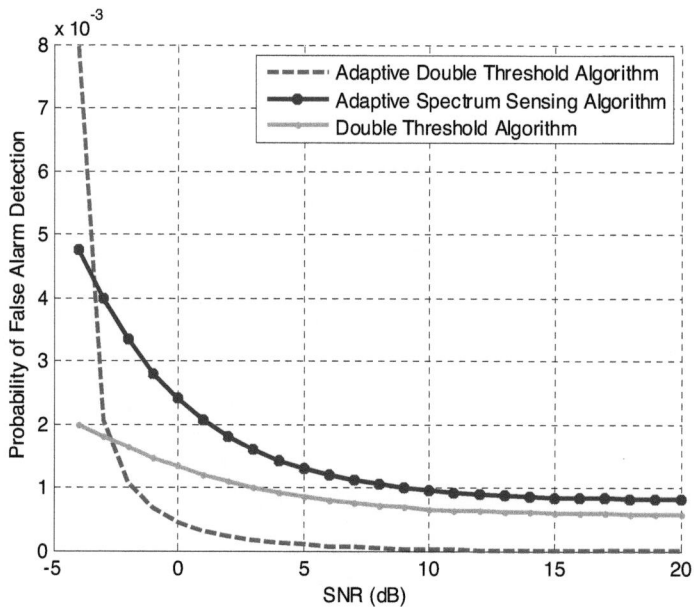

Figure 5.4: Plot of probability of false alarm for different values of SNR

Figure 5-4 shows the comparison of probability of false alarm with three algorithms. Throughout the range of SNR values, the probability of false alarm for Adaptive Double Threshold Spectrum Sensing is low as compared to others. This is as expected. Low probability of false alarm means the channel is best used by the secondary user. This increases the spectrum efficiency or the throughput of the secondary user network is enhanced. Therefore, we come with the conclusion that the Adaptive Double Threshold spectrum Sensing outwit the other two algorithms as far as the secondary user network throughput is taken into consideration.

Figure 5.5 shows the comparison of spectrum unavailability for three algorithms. It shows that the spectrum unavailable problem can be best observed in case of Double

Threshold Spectrum Sensing for low values of SNR. This problem is remarkably decreased in case of Adaptive Spectrum Sensing and further is lowered in Adaptive Double Threshold Spectrum Sensing. This implies that the performance of the secondary user can be increased and the interference which it can cause at lower SNR values to the primary user is lowered.

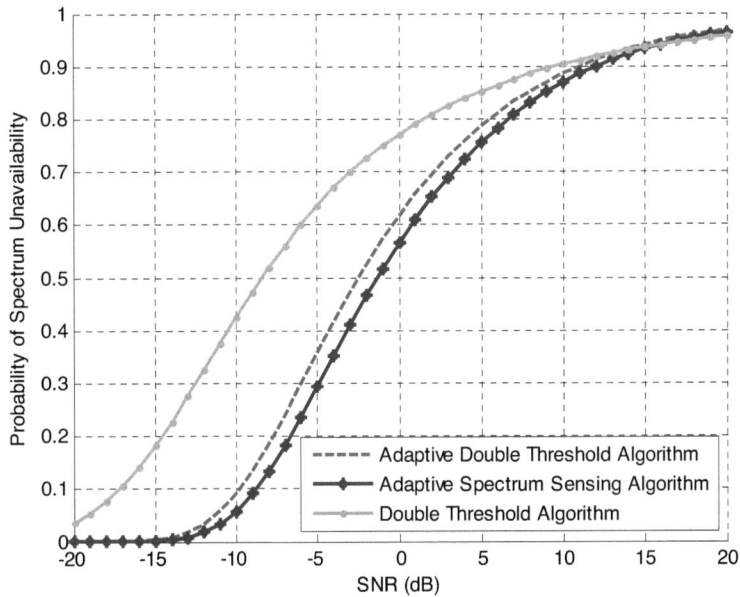

Figure 5.5: Plot of Spectrum Unavailability for different values of SNR

Figure 5.6 shows the comparison of the probabilities of collision between the cognitive user and the primary user changing with different values of SNR among the three algorithms. The simulation parameters are assumed as follows: the signal variance power = 1, the SNR values range from -20dB to 20dB, channel is AWGN, QPSK modulation with modulation order of 4, sample size = 10000 or above The figure shows that the collision probability is higher and reach upto unity for low SNR values (below -20dB, the case is worst) for both Adaptive Spectrum Sensing Algorithm and Adaptive Double Threshold Algorithm. The rate of decrement is sharp for increasing SNR. At the same time it is better in the case of Double Threshold Algorithm, however, its rate of decrement is slow for increasing SNR. For the higher values of SNR and above 10 dB, the collision probability is decreasing and reaching almost zero. Its value is found decreasing slowly in the case of Double Threshold Algorithm. The Adaptive Double Threshold is best among the three algorithms for higher SNR values (typically above -5dB). From the result we can draw the conclusion that the noise is uncertain in low SNR and this cause to increase the collision probability. The simulations were done for different values of SNR. The performance indicators and their dependencies on SNR for different algorithms were analyzed through simulations. The simulations were carried out with objective of increasing the performance of energy detection based spectrum sensing.

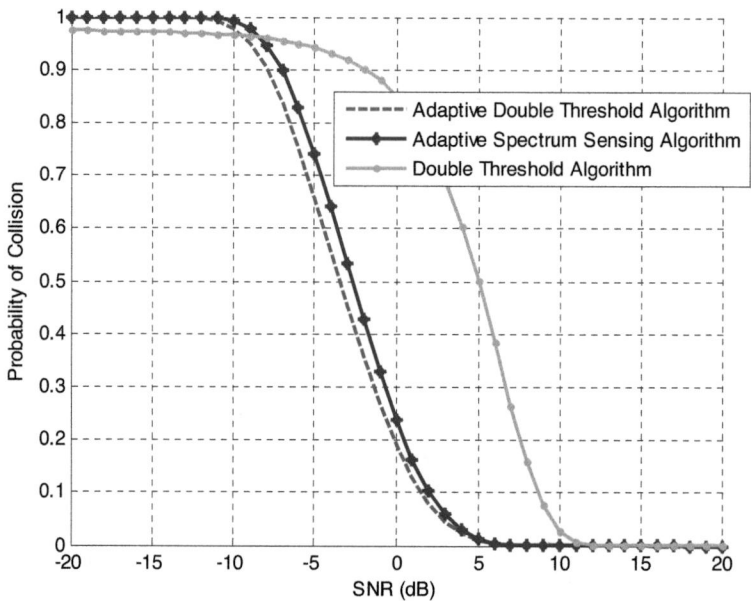

Figure 5.6: Plot of Probability of Collision for different values of SNR

Table 5-3: Comparative results of three Algorithms for SNR greater than -5dB

Energy Detection Algorithms	Probability of Detection	Probability of False Alarm	Probability of Spectrum Unavailability	Probability of Collision
Double Threshold Algorithm	Low	Medium	High	High
Adaptive Spectrum Sensing Algorithm	Medium	High	Medium	Medium
Adaptive Double Threshold Algorithm	High	Low	Low	Low

The concluding remarks of the simulation results are summarized in the above Table 4.3. The curves of collision probability and Spectrum Unavailability probability are compared for different algorithms. The curves showed their tradeoff. The curves of detection probability and false alarm probability are compared. Again, curves showed their tradeoff. The comparison results suggested that the performance of energy detection based spectrum sensing based on adaptive double threshold algorithm is better for lower values of SNR where noise power uncertainty is maximum. The adaptive double threshold energy detection works well where noise is uncertain based on the probability of detection and probability of false alarm. This suggests that the spectral efficiency is increased and interference caused to the primary user is also decreased. Therefore, the adaptive double threshold is better than the other when the noise is uncertain.

6 Conclusion and Recommendation

6.1 Conclusion

General survey of the energy based spectrum sensing based on different algorithms is done both in cooperative and non-cooperative nature of the secondary users. The performance of the energy based non - cooperative spectrum sensing is evaluated for different algorithms such as single threshold energy detection algorithm, double threshold energy detection algorithm, adaptive spectrum sensing algorithm and an adaptive double threshold spectrum sensing algorithm. The major conclusions drawn from the simulations are pointed out as

i) Conventional single threshold performs poor based on probability of detection for lower value of SNR.
ii) Double threshold energy detection performs poor based on probability of detection and probability of false alarm when noise is uncertain.
iii) The adaptive spectrum sensing algorithm optimizes the detection threshold of the energy detector when noise is uncertain.
iv) An adaptive double threshold spectrum sensing algorithm is better in performance than adaptive spectrum sensing algorithm, in case of detection probability would decline when SNR decreases under noise uncertainty.
v) Finally, the performance based on collision probability of adaptive double threshold algorithm is comparable to the adaptive spectrum sensing algorithm at extreme low SNR values typically below -5dB. However, the result is not satisfactory.

6.2 Recommendation

The Adaptive Double Threshold energy detector proved itself the efficient in case when noise is uncertain. But, its performance at low values of SNR needs to be enhanced further. This algorithm can be recommended to be used for the following scenarios:

i) The Adaptive double Threshold Energy detector can be incorporated in cooperative spectrum sensing for better performance.
ii) It can be the part of Hybrid spectrum Sensing techniques like any two combinations of spectrum sensing techniques. For example it can be integrate with matched detector or Cyclostationary detector both for non- cooperative and cooperative spectrum sensing for better performance.
iii) This algorithm can be applied for different fading channels through which its performance is evaluated both in cooperative and non – cooperative spectrum sensing

References

[1] Mitola, J. and J. Maguire, G.Q., "Cognitive radio: making software radios more personnel", IEEE personal commun. Mag; vol. 6, no. 4, pp. 13-18, Aug, 1999

[2] J. Mitola, Cognitive Radio: An integrated Agent Architecture for Software Defined Radio. PhD thesis, Royal institute of technology (KTH), 2000

[3] S. Haykin, "Cognitive radio: Brain – empowered wireless communication," IEEE Journal on Selected Areas in communication, vol. 23, pp. 201 – 220, February 2005.

[4] C. Ramming, "Cognitive networks," in DARPA Tech, 2004.

[5] Federal Communications Commission, "Notice of proposed rulemaking on cognitive radio," Tech. Rep. FCC 03 – 322, FCC, 2003.

[6] Yucek and Arslan, "A survey of spectrum sensing Algorithm for Cognitive Radio Applications," IEEE Communications Surveys and Tutorials, vol. 11, pp. 116 – 130, No. 1, First Quarter 2009.

[7] Yiping Xing; Mathur, C.N.; Haleem, M.A.; Chandramouli, R.; Subbalakshmi, K. P., "Dynamic Spectrum Access with QoS and Interference Temperature Constraints,"Mobile Computing, IEEE Transactions on , vol.6, no.4, pp.423,433, April 2007

[8] Beibei Wang; Liu, K.J.R., "Advances in cognitive radio networks: A survey," Selected Topics in Signal Processing, IEEE Journal of , vol.5, no.1, pp.5,23, Feb. 2011

[9] Tandra, R.; Sahai, A., "Fundamental limits on detection in low SNR under noise uncertainty," Wireless Networks, Communications and Mobile Computing, 2005 International Conference on , vol.1, no., pp.464,469 vol.1, 13-16 June 2005

[10] Olivieri, M.P.; Barnett, G.; Lackpour, A.; Davis, A.; Ngo, P., "A scalable dynamic spectrum allocation system with interference mitigation for teams of spectrally agile software defined radios," New Frontiers in Dynamic Spectrum Access Networks, 2005. DySPAN 2005. 2005 First IEEE International Symposium on , vol., no., pp.170,179, 8-11 Nov. 2005

[11] Weidling, F.; Datla, D.; Petty, V.; Krishnan, P.; Minden, G.J., "A framework for R.F. spectrum measurements and analysis," New Frontiers in Dynamic Spectrum Access Networks, 2005. DySPAN 2005. 2005 First IEEE International Symposium on , vol., no., pp.573,576, 8-11 Nov. 2005

[12] D.-C. Oh and Y.-H. Lee, "Energy detection based spectrum sensing for sensing error minimization in cognitive radio networks," Int. J. Commun. Netw. Inf. Security (IJCNIS), vol. 1, no. 1, Apr. 2009

[13] Vartiainen, J.; Sarvanko, H.; Lehtomaki, J.; Juntti, M.; Latva-aho, M., "Spectrum Sensing with LAD-Based Methods," Personal, Indoor and Mobile Radio

Communications, 2007. PIMRC 2007. IEEE 18th International Symposium on , vol., no., pp.1,5, 3-7 Sept. 2007

[14] Wylie-Green, M.P., "Dynamic spectrum sensing by multiband OFDM radio for interference mitigation," New Frontiers in Dynamic Spectrum Access Networks, 2005. DySPAN 2005. 2005 First IEEE International Symposium on , vol., no., pp.619,625, 8-11 Nov. 2005

[15] Tandra, R.; Sahai, A., "SNR Walls for Signal Detection," Selected Topics in Signal Processing, IEEE Journal of , vol.2, no.1, pp.4,17, Feb. 2008

[16] Ruiliang Chen; Jung-Min Park, "Ensuring Trustworthy Spectrum Sensing in Cognitive Radio Networks," Networking Technologies for Software Defined Radio Networks, 2006. SDR '06.1st IEEE Workshop on , vol., no., pp.110,119, 25-25 Sept. 2006

[17] Burbank, J.L., "Security in Cognitive Radio Networks: The Required Evolution in Approaches to Wireless Network Security," Cognitive Radio Oriented Wireless Networks and Communications, 2008. CrownCom 2008. 3rd International Conference on , vol., no., pp.1,7, 15-17 May 2008

[18] Wild, B.; Ramchandran, K., "Detecting primary receivers for cognitive radio applications," New Frontiers in Dynamic Spectrum Access Networks, 2005. DySPAN 2005. 2005 First IEEE International Symposium on , vol., no., pp.124,130, 8-11 Nov. 2005

[19] Brown, T.X., "An analysis of unlicensed device operation in licensed broadcast service bands," New Frontiers in Dynamic Spectrum Access Networks, 2005. DySPAN 2005. 2005 First IEEE International Symposium on , vol., no., pp.11,29, 8-11 Nov. 2005

[20] Ian F. Akyildiz, Won-Yeol Lee, Mehmet C. Vuran, ShantidevMohanty,NeXt generation/dynamic spectrum access/cognitive radio wireless networks: A survey, Computer Networks, Volume 50, Issue 13, 15 September 2006, Pages 2127-2159, ISSN 1389-1286, 10.1016/j.comnet.2006.05.001.

[21] Garello, R.; Jia, Y., "Comparison of spectrum sensing methods for cognitive radio under low SNR," Antennas and Propagation in Wireless Communications (APWC), 2011 IEEE-APS Topical Conference on , vol., no., pp.886,889, 12-16 Sept. 2011

[22] Digham, F.F.; Alouini, M.-S.; Simon, Marvin K., "On the energy detection of unknown signals over fading channels," Communications, 2003. ICC '03. IEEE International Conference on , vol.5, no., pp.3575,3579 vol.5, 11-15 May 2003

[23] Jinbo Wu; Tao Luo; GuangxinYue, "An Energy Detection Algorithm Based on Double-Threshold in Cognitive Radio Systems," Information Science and Engineering (ICISE), 2009 1st International Conference on , vol., no., pp.493,496, 26-28 Dec. 2009

[24] Bin Shen; Longyang Huang; Chengshi Zhao; Zheng Zhou; KyungsupKwak, "Energy Detection Based Spectrum Sensing for Cognitive Radios in Noise of

48

Uncertain Power," Communications and Information Technologies, 2008. ISCIT 2008. International Symposium on , vol., no., pp.628,633, 21-23 Oct. 2008

[25] Jiang Zhu; ZhengguangXu; Furong Wang; Benxiong Huang; Bo Zhang, "Double Threshold Energy Detection of Cooperative Spectrum Sensing in Cognitive Radio," Cognitive Radio Oriented Wireless Networks and Communications, 2008. CrownCom 2008. 3rd International Conference on , vol., no., pp.1,5, 15-17 May 2008

[26] Reisi, N.; Ahmadian, M.; Salari, S., "Performance Analysis of Energy Detection-Based Spectrum Sensing over Fading Channels," Wireless Communications Networking and Mobile Computing (WiCOM), 2010 6th International Conference on , vol., no., pp.1,4, 23-25 Sept. 2010

[27] Shibing Zhang; ZhihuaBao, "An Adaptive Spectrum Sensing Algorithm under Noise Uncertainty," Communications (ICC), 2011 IEEE International Conference on , vol., no., pp.1,5, 5-9 June 2011

[28] JinquanXie; Jin Chen, "An Adaptive Double-Threshold Spectrum Sensing Algorithm under Noise Uncertainty," Computer and Information Technology (CIT), 2012 IEEE 12th International Conference on , vol., no., pp.824,827, 27-29 Oct. 2012

49